OECD Environmental Performance Reviews

Biodiversity Conservation and Sustainable Use in Latin America

EVIDENCE FROM ENVIRONMENTAL PERFORMANCE REVIEWS

This work is published under the responsibility of the Secretary-General of the OECD. The opinions expressed and arguments employed herein do not necessarily reflect the official views of OECD member countries.

This document, as well as any data and any map included herein, are without prejudice to the status of or sovereignty over any territory, to the delimitation of international frontiers and boundaries and to the name of any territory, city or area.

Please cite this publication as:
OECD (2018), *Biodiversity Conservation and Sustainable Use in Latin America: Evidence from Environmental Performance Reviews*, OECD Environmental Performance Reviews, OECD Publishing, Paris.
https://doi.org/10.1787/9789264309630-en

ISBN 978-92-64-30960-9 (print)
ISBN 978-92-64-30963-0 (pdf)

Series: OECD Environmental Performance Reviews
ISSN 1990-0104 (print)
ISSN 1990-0090 (online)

The statistical data for Israel are supplied by and under the responsibility of the relevant Israeli authorities. The use of such data by the OECD is without prejudice to the status of the Golan Heights, East Jerusalem and Israeli settlements in the West Bank under the terms of international law.

Photo credits: Cover © Shutterstock.com/Gustavo Frazao; © Shutterstock.com/BonnieBC.

Corrigenda to OECD publications may be found on line at: *www.oecd.org/publishing/corrigenda*.
© OECD 2018

You can copy, download or print OECD content for your own use, and you can include excerpts from OECD publications, databases and multimedia products in your own documents, presentations, blogs, websites and teaching materials, provided that suitable acknowledgement of OECD as source and copyright owner is given. All requests for public or commercial use and translation rights should be submitted to *rights@oecd.org*. Requests for permission to photocopy portions of this material for public or commercial use shall be addressed directly to the Copyright Clearance Center (CCC) at *info@copyright.com* or the Centre français d'exploitation du droit de copie (CFC) at *contact@cfcopies.com*.

Preface

Latin America is one of the most important regions of the world in terms of biodiversity and ecosystems. The region's rich biodiversity provides invaluable benefits to human health, well-being and the broader economy. However, large-scale deforestation to clear land for agriculture, mining, energy and infrastructure projects, over-extraction of natural resources, invasive species and climate change are placing enormous pressure on the region's natural wealth.

This report summarises key findings and lessons learned in the area of biodiversity conservation and sustainable use from the OECD Environmental Performance Reviews conducted for five Latin American countries between 2013 and 2017: Brazil, Chile, Colombia, Mexico and Peru.

The report highlights the leadership of the region in the use of payments for ecosystem services. It shows that terrestrial and marine protected areas in some Latin American countries cover a surface far surpassing the international Aichi Targets of 17% and 10% respectively by 2020. Innovative solutions such as conservation trust funds are increasingly used to bridge the large biodiversity finance gap. The report also points to the particularly high potential for nature based tourism in Latin America, especially in coastal and marine ecosystems.

The report describes remaining challenges that need to be addressed. The rate of deforestation in South America has slowed down, but remains among the highest in the world. Economic instruments such as charges and fees for pollution are still in the early stages of development. Despite growing water scarcity risks in the region, irrigation practices have yet to significantly shift to modern water saving practices. Environmentally harmful subsidies continue to provide perverse incentives for biodiversity conservation and sustainable use. Agricultural support systems have yet to be reformed to discourage pesticide use.

The report builds on the wealth of policy analysis provided in the Environmental Performance Reviews. It is the result of a constructive dialogue between the OECD and its member and partner countries participating in the OECD Working Party of Environmental Performance. I am confident that this effort will be helpful to improve our understanding of the challenges for biodiversity conservation and sustainable use and to identify good practices and innovative solutions to improve the management of biodiversity in Latin America and beyond.

Rodolfo Lacy
Director, OECD Environment Directorate

Foreword

The OECD Environmental Performance Review Programme has been supporting member and partner countries in developing effective environmental policies for nearly 30 years. The principal aim of the programme is to:

- support countries evaluate progress in achieving their environmental goals;
- promote continuous policy dialogue and peer learning; and
- stimulate greater accountability from governments towards each other and public opinion.

The thematic reports developed under the Environmental Performance Review Programme contribute to these objectives by summarising experience and lessons learned from country specific Environmental Performance Reviews.

This report provides a cross-country overview of biodiversity conservation and sustainable use policies in five Latin American countries. It draws on the OECD's Environmental Performance Reviews completed for Brazil, Chile, Colombia, Mexico and Peru between 2013 and 2017. It presents the main challenges facing these Latin American countries, the strategies being used to tackle them and the gaps that remain. The report takes into account major policy changes since these countries' Environmental Performance Reviews depending on information availability, and presents updated data and indicators across selected Latin American countries. However, as the countries were reviewed over several years, information for some countries may be more recent than others. Nevertheless, the policy recommendations emerging from the reviews may provide useful lessons for other OECD and partner countries.

The authors of the report are Anna Drutschinin and Britta Labuhn from the OECD Environment Directorate and Rachel Samson of Carist Consulting. Ivana Capozza of the OECD Environment Directorate co-ordinated its preparation. Nathalie Girouard, Head of the Environmental Performance and Information Division of the OECD Environment Directorate provided oversight and guidance. Preparation of the report benefited from the contribution of Environment Directorate colleagues including Jane Ellis and Katia Karousakis, as well as from consultation with the OECD Working Party on Environmental Performance and the Working Party on Biodiversity, Water and Ecosystems. Carla Bertuzzi provided statistical support and Mika Hosokawa, Annette Hardcastle and Natasha Cline-Thomas prepared the report for publication.

The figures presented in this report are based on data available up to October 2018.

Table of contents

Preface 3
Foreword 5
Abbreviations and acronyms 10
Executive summary 11
  Institutional and policy frameworks 11
  Policy instruments 11
  Financing 11
  Mainstreaming biodiversity 11

1. Main findings and conclusions 13
  1.1. Introduction 14
  1.2. Status, trends and pressure on biodiversity 14
  1.3. Governance and policy framework 15
  1.4. Knowledge base and evaluation 16
  1.5. Policy instruments 17
  1.6. Financing 18
  1.7. Mainstreaming 19
  Note 20

2. Trends and key pressures on biodiversity and ecosystems 21
  2.1. Status and trends 22
  2.2. Key pressures 26
  References 29

3. Institutional and policy frameworks 31
  3.1. Introduction 32
  3.2. Governance and institutions 32
  3.3. Stakeholder participation and engagement of indigenous peoples and traditional communities 34
  3.4. Biodiversity strategies and legislation 35
  3.5. International and regional co-operation 36
  3.6. Status of data and knowledge 37
  References 40

4. Policy instruments 41
  4.1. Introduction 42
  4.2. Regulatory instruments 43
  4.3. Economic instruments 48
  4.4. Voluntary and information instruments 55
  References 56

5. Financing 57
   5.1. Introduction 58
   5.2. Domestic public financing 58
   5.3. Private revenues 60
   5.4. Biodiversity funds 61
   5.5. International financing 63
   References 64

6. Mainstreaming 67
   6.1. Introduction 68
   6.2. Agriculture 69
   6.3. Fishing and aquaculture 72
   6.4. Forestry 74
   6.5. Energy and infrastructure 79
   6.6. Tourism 80
   6.7. Climate change 81
   Notes 82
   References 82

Tables

Table 4.1. Policy instruments for biodiversity in Latin America 42

Figures

Figure 1.1. Forest loss and the number of threatened species are high 15
Figure 1.2. Protected areas are the predominant policy instrument used for biodiversity conservation 17
Figure 2.1. Forest loss remains high 24
Figure 2.2. Ocean Health Index scores are deteriorating 25
Figure 2.3. The number of threatened species is high 26
Figure 2.4. Invasive species are a threat to biodiversity 28
Figure 4.1. Several countries exceed the CBD target for terrestrial areas 44
Figure 5.1. Protected area funding is uneven 59
Figure 5.2. Biodiversity-related ODA commitments are significant 64
Figure 6.1. Most Latin American countries are increasing their pesticide use 70
Figure 6.2. Agriculture is the largest water user in most countries 71
Figure 6.3. Fish catches are declining, while aquaculture is growing 73
Figure 6.4. Certified forest still represents a relatively small proportion of total forest area ... 75
Figure 6.5. Mining is an important part of many Latin American economies 77

Boxes

Box 2.1. Biodiversity hotspots of Latin America 22
Box 3.1. Mexico's Inter-Ministerial Commission on Climate Change 33
Box 3.2. Colombia's approach to access and benefit sharing 37
Box 3.3. Brazil's deforestation monitoring systems 38
Box 3.4. Economic valuation of biodiversity 39
Box 4.1. Amazon Region Protected Areas programme 45
Box 4.2. Brazil improves governance and management of protected areas 46

Box 4.3. Chile's Pumalin Park, one of the largest private protected areas 47
Box 4.4. Linking PES with social protection: Bolsa Verde and Bolsa Floresta 49
Box 4.5. Chingaza National Park in Colombia values ecosystem services from the Páramo 50
Box 5.1. Peru increases financing for environmental enforcement 60
Box 5.2. The Brazilian Amazon Fund 62
Box 6.1. Irrigation practices in Latin America 72
Box 6.2. Slowing deforestation in Brazil: Progress and challenges 76
Box 6.3. Chile's mine closure financial guarantees 79

Follow OECD Publications on:

 http://twitter.com/OECD_Pubs

http://www.facebook.com/OECDPublications

 http://www.linkedin.com/groups/OECD-Publications-4645871

http://www.youtube.com/oecdilibrary

 http://www.oecd.org/oecddirect/

This book has...

A service that delivers Excel® files from the printed page!

Look for the *StatLinks* at the bottom of the tables or graphs in this book. To download the matching Excel® spreadsheet, just type the link into your Internet browser, starting with the *http://dx.doi.org* prefix, or click on the link from the e-book edition.

Abbreviations and acronyms

ARPA	Amazon Region Protected Areas
Biofin	UNDP's Biodiversity Finance Initiative
CBD	Convention on Biological Diversity
CONABIO	Mexico's National Commission for the Knowledge and Use of Biodiversity
ECLAC	Economic Commission for Latin America and the Caribbean
EIA	Environmental impact assessment
EPR	Environmental Performance Review
Funbio	Brazilian Biodiversity Fund
GHG	Greenhouse gas
IUCN	International Union for the Conservation of Nature
NGO	Non-governmental organisation
ODA	Official development assistance
PES	Payment for ecosystem services
SEA	Strategic environmental assessment
WAVES	World Bank Wealth Accounting and the Valuation of Ecosystem Services
ZEE	Environmental and ecological-economic zoning

Executive summary

Latin America is one of the most important regions of the world in terms of biodiversity and ecosystems. The region's rich biodiversity provides invaluable benefits to human health, well-being and the broader economy. However, its wealth is under threat. As in other parts of the world, large-scale deforestation to clear land for agriculture, mining, energy and infrastructure projects are placing enormous pressure on the region's ecosystems. Invasive alien species, overfishing and climate change are additional drivers of biodiversity loss.

Institutional and policy frameworks

Institutional and policy frameworks for biodiversity conservation and sustainable use in Latin America have improved. International agreements such as the UN Convention on Biological Diversity have triggered the revision and adoption of new strategies and action plans. The diversity of policy instruments used has increased. However, lack of financial and human capacity, poor co-ordination, low political priority assigned to biodiversity, insufficient data, and inadequate mainstreaming of biodiversity considerations into sectoral policies continue to hamper effective implementation.

Policy instruments

Protected areas are the main biodiversity conservation tool used in Latin America. Terrestrial protected areas cover a surface far surpassing the international Aichi Target of 17% by 2020, but marine protected areas lag behind. The region is a leader in the use of payments for ecosystem services, yet the use of other economic instruments – such as water charges, water markets, forestry fees and tradable fishing and forestry quotas – could be further extended. Environmentally harmful subsidies, such as for pesticides and small-scale mining, are impeding progress and need reform.

Financing

Government budgets for biodiversity have been increasing, and are complemented by international development finance. However, the overall level of funding remains inadequate. Increases need to come both from public budgets and from external sources, which governments could leverage through the greater use of economic instruments and public-private partnerships. More work could be done to ensure that finance is channelled to where it is needed most, which will require improvements in data and information.

Mainstreaming biodiversity

Aligning sectoral and biodiversity objectives are important, as development continues to put pressure on areas outside of official protection. Effective mainstreaming requires good governance and political and financial engagement. While the use of environmental impact assessments or strategic environmental assessment is growing, work is required to ensure that the approaches are accepted by local communities. Mainstreaming is important in agriculture, forestry, fisheries, tourism and mining, energy and infrastructure development as these sectors depend on natural resources and the services that ecosystems provide, but also have negative effects on biodiversity. Synergies between biodiversity and climate change mitigation and adaptation are being explored and should be further capitalised.

1. Main findings and conclusions

This chapter summarises the report's main findings and conclusions. It begins with an overview of the main status, trends and pressures on biodiversity and ecosystems in the reviewed Latin American countries, followed by a discussion on the institutional and policy frameworks, financing, and mechanisms to mainstream biodiversity into sectoral policies. For each section, the chapter highlights some good practices and innovative solutions implemented in the countries.

The statistical data for Israel are supplied by and under the responsibility of the relevant Israeli authorities. The use of such data by the OECD is without prejudice to the status of the Golan Heights, East Jerusalem and Israeli settlements in the West Bank under the terms of international law.

1.1. Introduction

Biodiversity and ecosystem services underpin human well-being and play a critical role in the economy. The OECD has reviewed policies for the conservation and sustainable use of biodiversity in more than a dozen countries in the framework of its Environmental Performance Reviews (EPRs) since 2010.[1] While pressures on ecosystems and biodiversity are diverse, the policy recommendations that emerged from the reviews may provide useful lessons for other OECD and partner countries.

This paper provides a cross-country review of policies for biodiversity conservation and sustainable use in Latin America, based on the EPRs of five countries in the region:

- Mexico (2013)
- Colombia (2014)
- Brazil (2015)
- Chile (2016)
- Peru (2017)

Note: Add the note here. If you do not need a note, please delete this line.
Source: Add the source here. If you do not need a source, please delete this line.

Focusing on Latin America is particularly pertinent given the great wealth of biodiversity in the region and the growing pressures on its conservation and sustainable use. This paper describes common trends, challenges and key achievements in the five examined countries. Where appropriate, it also brings in evidence from other Latin American countries. As the paper draws on EPRs published over the past five years, information for some countries may be more recent than for others.

1.2. Status, trends and pressure on biodiversity

Latin America and the Caribbean is one of the most important regions of the world in terms of biodiversity and ecosystems, holding an estimated 40% of the world's biological diversity. Six of the 17 "megadiverse countries" are within Latin America and the region hosts 11 of the 14 terrestrial biomes, about half of global forests, and the second largest reef system in the world. However, high economic and population growth are driving land-use change, creating pollution and increasing resource demand.

Forest loss remains one of the greatest pressures on biodiversity and ecosystems in Latin America. The rates of annual forest loss have generally declined over the past years, but they remain among the highest in the world. Brazil in particular significantly slowed its deforestation rate while some countries expanded their forest area (Figure 1.1). Water resources are abundant, but many arid and semi-arid regions are facing increasing water scarcity as a result of growing water demand and reduced water availability due to climate change. Latin American countries have high proportions of endemic species (that are found nowhere else in the world), but they also have some of the highest numbers of threatened species in the world. Biodiversity "hotspots", which combine high degrees of endemism and biodiversity loss, extend across many countries in the region.

Land clearing for agriculture is the largest cause for forest loss; and is often exacerbated by unclear or lack of land tenure. Illegal activities such as mining, traffic of species, timber harvesting and illicit crops are a particular problem in the region (e.g. in Colombia and Peru). Agriculture is another significant threat to biodiversity, as a result of overgrazing, pesticide and fertiliser use, and high water use. Mining and energy extraction and infrastructure are also important drivers of biodiversity loss, due the land-use change, high groundwater extraction, soil and water contamination and the hazardous waste generation they often involve. The impacts of these pressures are not always well-assessed and mitigated. Invasive species are of particular concern: some 54 of the world's 100 worst invasive species are present in the region. Marine and coastal ecosystems are threatened by expanding coastal development, overfishing, bycatch, pollution, untreated waste, unsustainable tourism and invasive species, while inland water systems are challenged by pollution and excessive water use. Latin American countries mainly rank below the global average in the 2017 Ocean Health Index assessment.

Figure 1.1. Forest loss and the number of threatened species are high

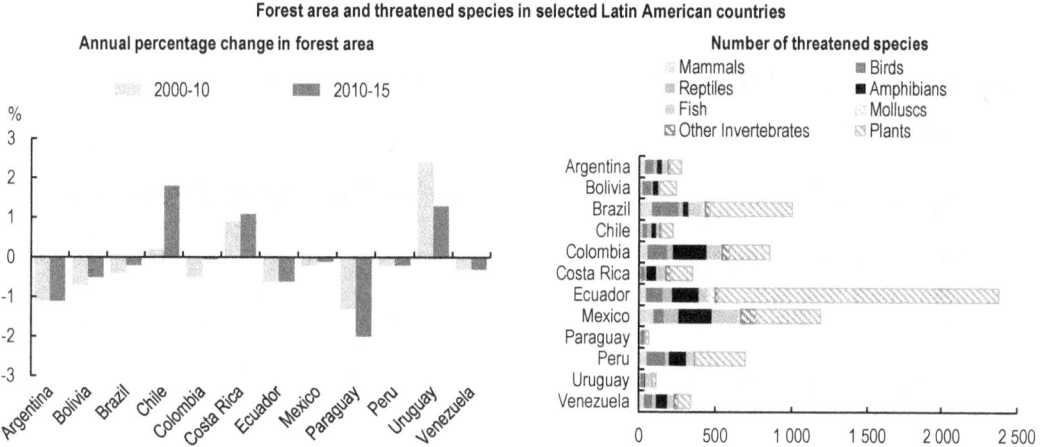

Source: FAO (2015), *Global Forest Resource Assessment 2015*, www.fao.org/forest-resources-assessment/en/; IUCN (2018), *IUCN Red List of Threatened Species*, www.iucnredlist.org/.

StatLink https://doi.org/10.1787/888933885980

1.3. Governance and policy framework

All five of the countries examined have improved their institutional frameworks and governance systems for biodiversity, with environment ministries leading biodiversity policy development and dedicated agencies implementing and managing protected areas in four of the five countries. Colombia and Peru have advanced efforts to decentralise environmental responsibilities to sub-national and local authorities. However, their experience has shown that the best results are achieved only when these authorities have sufficient human and financial resources to fulfil their responsibilities, which is not always the case at present. Better co-ordination and targeted support to regions and municipalities most in need of strengthening in their technical and financing capacities would support progress across sub-national jurisdictions.

There is a greater prevalence of consultations, public hearings, and inclusion of stakeholders in environmental management councils, as well as more environmental courts and tribunals to address cases of environmental conflict. A lot of work also remains to rebuild trust with communities regarding their involvement in decision-making processes, in order to reduce environmental conflict.

International agreements, notably the UN Convention on Biological Diversity (CBD), are helping to drive further domestic action, leading countries to update and revise their national biodiversity strategies to incorporate the 2011-20 Aichi Targets under the CBD. Much progress has been made in developing legislation, goals and targets for biodiversity conservation, although implementation continues to be a challenge as a result of a lack of resources, capacity, co-ordination and political leadership. Recent strategies tend to put emphasis on enabling factors, such as knowledge creation, capacity building and awareness rising. Ecosystem restoration and connectivity, establishment of priority conservation regions, synergies amongst biodiversity-related conventions and biodiversity mainstreaming with sectoral policies is also gaining in importance. Interest in the sustainable use of biodiversity, including biotechnology from generic resources, is increasing.

Several countries have embraced the 2014 Nagoya Protocol on Access and Benefit Sharing, although most still lack legal or regulatory frameworks to govern the access to genetic resources and ensure fair and equitable sharing of benefits arising from their use. Cross-country collaboration has borne fruit and could be further pursued to pool and leverage knowledge, resources and capacity. Regional initiatives, such as the Latin American Initiative for Sustainable Development, are facilitating improved information sharing, policy co-ordination and harmonisation. Additionally, many Latin American countries have bilateral or multilateral co-operation agreements that address shared ecosystems or provide financing and capacity building for biodiversity conservation efforts.

> **Mexico** has a National Commission for the Knowledge and Use of Biodiversity (CONABIO), with representation from ten ministries as well as consultative bodies to facilitate public participation in biodiversity matters. **Brazil** shifted from a "fence-and-protect" approach to one that favours the sustainable use of biological resources and recognises the role of rural, traditional and indigenous communities in preserving ecosystems. **Colombia** has integrated biodiversity into its National Development Plan and released the National Policy for the Integral Management of Biodiversity and Ecosystem Services (PNGIBSE) in 2012.

1.4. Knowledge base and evaluation

The lack of knowledge and data remain a key challenge, particularly concerning marine and freshwater ecosystems despite improvement in the breadth and depth of environmental indicators over the past decade. Knowledge and data on the status of ecosystems and species, monitoring and reporting of trends, and insight into the economic and social value of biodiversity are essential for building awareness, establishing priorities for action and effectively designing and implementing biodiversity policy. There are, however, some positive examples that could be usefully replicated in other countries.

> **Mexico** has a highly developed biodiversity information system and has conducted several economic valuation studies related to biodiversity and ecosystem services. **Brazil** has world-leading satellite-based deforestation monitoring systems, which have been crucial to reducing forest clearing in the Amazon. **Colombia** has conducted an independent assessment of its marine ecosystems using the international Ocean Health Index. It also participates in the World Bank Wealth Accounting and the Valuation of Ecosystem Services (WAVES) initiative, developing natural capital accounting for its water, forests and land.

1.5. Policy instruments

Protected areas are the predominant policy instrument used for biodiversity conservation in Latin America. Terrestrial protection covered 28% of Central America and 25% of South America in 2014, the largest shares in the world and far surpassing the area protected in OECD countries (15%) as well as the international target of 17% under the CBD. However, quality is just as important as quantity, and it is here where the five EPR countries could improve in ensuring that all biomes and ecosystems are represented and adequately resourced to be managed effectively. Marine protection is much lower than terrestrial protection (2% and 4% of total marine areas in Central and South America respectively) and needs to be expanded. The private sector can help expand and finance protected areas, for example through public-private partnerships in areas with high tourism potential, or through incentives encouraging environmental investment and philanthropy, for example in the form of financial and land donations.

Figure 1.2. Protected areas are the predominant policy instrument used for biodiversity conservation

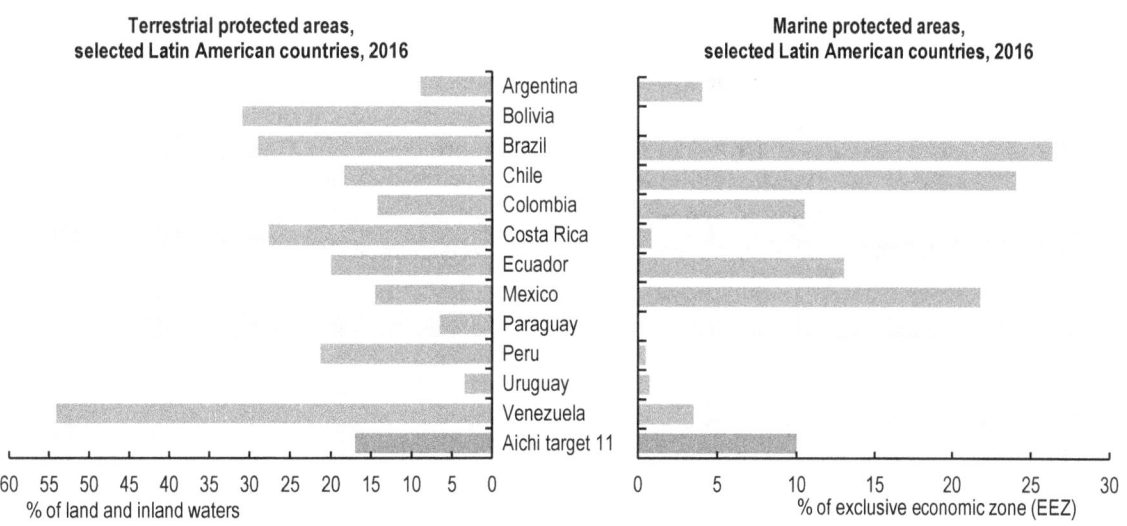

Note: Data for Chile include the largest marine reserve in the Americas (Nazca-Desventuradas). Data for Brazil include two large mosaics of marine protected areas designated in March 2018 (Archipelago of Trindade and Martim Vaz and Monte Columbia and Archipelago of São Pedro and São Paulo).
Source: UNEP-WCMC and IUCN (2018), *The World Database on Protected Areas* (WDPA), www.protectedplanet.net.

StatLink https://doi.org/10.1787/888933885999

The usage of economic instruments for biodiversity conservation and sustainable use is growing in Latin America. The region is a leader in payments for ecosystem services (PES). There are several large-scale programmes, some of which have interesting features, focussing for example on areas with high biodiversity benefits, high risk of loss, low opportunity cost, or low social development. Several countries introduced national legislation to facilitate the use of PES, yet there is a general lack of monitoring frameworks that would allow for evaluating the effectiveness of programmes.

Offsets, water charges, water markets, forestry fees and tradable fishing and forestry quotas could be further extended. Environmentally harmful subsidies are impeding progress, such as tax exemptions for fertilisers and pesticides and subsidies for irrigation infrastructure and small-scale mining. Phasing these out would have the double benefit of stopping the promotion of practices harmful to biodiversity and increasing public finances.

Voluntary and information-based instruments are increasingly being used as tools that integrate economic and environmental objectives, and have the potential to be used more widely. The number of environmental certifications and labels is growing, in part because of growing global demand for more sustainable products. Green public procurement initiatives, such as Brazil's programme targeting biodiversity-related products and Mexico's incorporation of certified forest products into its green procurement criteria, are encouraging growth of sectors that use resources sustainably. The "Soya Moratorium", involving a group of large companies that stopped buying soya grown on deforested land in the Brazilian Amazon, showed that private sector agreements can have significant positive impacts.

> **Peru** has signed at least ten administration contracts with non-government institutions to implement management plans in protected areas. **Mexico** runs one of the world's largest payment for ecosystem services (PES) programmes and **Brazil** has combined its PES programmes with social objectives, providing payments to extremely poor households in rural forest communities to compensate them for conservation activities. Both countries are using biodiversity offsets to help landowners comply with legal land conservation requirements. **Colombia** charges hydroelectric and thermal energy plants to carry out watershed conservation projects while Chile uses a market to allocate water abstraction rights. **Brazil** and **Chile** have concluded voluntary agreements with the private sector to help combat deforestation and to improve efficiency and sustainability of industry, respectively.

1.6. Financing

Financing biodiversity conservation and sustainable use initiatives remains one of the most significant challenges for Latin America. In the face of competing priorities and limited financial resources, biodiversity initiatives are often not fully funded. Protected area management, enforcement of existing environmental laws and ecosystem monitoring and reporting are particularly impacted by the lack of financial resources.

All five countries have increased their public budgets for biodiversity. Peru stands out, having increased its budget more than four-fold between 2012 and 2015. Several countries have created biodiversity funds to pool international and private finance and thereby facilitate investment in conservation initiatives. The region is a major recipient of international biodiversity finance: in 2011-15, seven of the top ten countries with the highest share of biodiversity-related finance in total official development assistance (ODA) were located in Latin America and the Caribbean. Nevertheless, finance needs to be scaled up from all sources, both public and private, for biodiversity objectives to be achieved. There is scope to increase the use of economic instruments to raise revenue, such as protected area entrance fees. Improving the efficiency of its use will also remain important.

> **Chile, Mexico, Brazil** and **Peru** all increased the engagement of the private sector to raise funds for protected area funding, particularly for areas with high tourism potential. All five examined countries are participating in UNDP's Biodiversity Finance Initiative (Biofin), which supports countries identify finance needs and gaps. **Brazil** is expanding the use of biodiversity funds to pool resources and allocate them more efficiently. The Brazilian Biodiversity Fund (Funbio), a non-profit private organisation, raises and invests financial resources for biodiversity conservation on behalf of the federal and state governments, in addition to large-scale conservation trust funds (such as the Amazon Fund).

1.7. Mainstreaming

Mainstreaming and aligning sectoral and biodiversity objectives are important in the region, as development continues and many areas remain outside official protection. Effective mainstreaming requires a number of framework conditions to be in place: good governance, effective processes, strong institutions, and political and financial investment and engagement. While improvements have been made – as shown for example by the growing use of strategic environmental assessment of policies, plans and programmes – significant further work is required to ensure that the approaches are comprehensive, consistent, effective and accepted by local communities. Mainstreaming is particularly important in agriculture, forestry, fishing and aquaculture, tourism, and mining, energy and infrastructure development as these sectors depend heavily on natural resources and ecosystems services. However, they are also key sources of ecosystem degradation and conflict. Synergies between biodiversity and climate change mitigation and adaptation are explored and could be further capitalised, for example by means of ecosystem-based approaches.

Reforming support systems is essential to integrate biodiversity into the agriculture sector. The use of pesticides and fertilisers continues to grow in many countries; and it remains subsidised in Mexico, for example. Instruments promoting harmful practices must be phased out, while programmes targeting rural poverty or other social issues are simultaneously introduced or expanded to minimise negative social ramifications. In the forestry sector, the use of certification and afforestation programmes has increased, and large-scale commercial forestry operators have greatly improved their performance. However, greater emphasis is needed on limiting loss of native forest and protecting and restoring priority areas for biodiversity. In both sectors, formalising land tenure, a significant challenge in the region, will be crucial for biodiversity mainstreaming to be a success. Quota systems have been introduced and regulations tightened in the fishing and aquaculture sector. However, monitoring and enforcement must be strengthened, for example concerning regulations to limit effluent, pesticides and medicines from fish farms.

Tourism offers Latin America significant economic opportunity and potential for increased biodiversity financing, but its expansion also presents a risk to biodiversity and ecosystems. The five EPR countries have increased their focus on nature-based tourism. Chile and Mexico have developed national strategies for sustainable tourism; and several countries are pursuing environmental certification schemes for the sector. These efforts are in their early stages and should be sustained.

> **Brazil** is greening its agricultural sector by making access to subsidised rural credit in the Amazon conditional on the legitimacy of land claims and compliance with environmental regulations. **Colombia** provides financial incentives for investment in eco-tourism and in forest plantations that favour native species over introduced species. It also has an ambitious strategy that calls for the return of 10 million ha of pasture and agricultural land to a more natural state, for example through reforestation. **Chile** established over hundreds of areas where exclusive rights are assigned to organisations of artisanal fishers, and amended its Law on Fishing and Aquaculture to base the establishment of its fishing quotas on scientific and technical factors. **Peru** has established a dedicated office to resolve mining-related disputes.

Mining is an important source of revenue in the five EPR countries, and energy and other infrastructure is expanding to meet the needs of growing economies and populations. Improvements have been made in legislative and regulatory processes, yet further effort is needed to enhance environmental monitoring and enforcement, ensure rigorous and collaborative environmental impact assessment processes, address conflicts with local and indigenous populations, and accelerate clean-up of abandoned mines. Environmental impact assessments often come too late in the process to significantly alter projects, lack consistent integration of biodiversity concerns, do not apply to smaller projects (e.g. small-scale and artisanal mining) or provide for only limited public participation in the process. More work is also needed to improve the application of land use planning, strategic environmental assessments, and other economic and environmental analysis surrounding infrastructure investment decisions.

Note

[1] These include the EPRs of Japan (2010), Norway, Israel (2011), Mexico, South Africa (2013), Colombia, Sweden (2014), Poland, Spain, Brazil (2015), France, Chile (2016), Switzerland (2017) and Hungary (2018).

2. Trends and key pressures on biodiversity and ecosystems

This chapter provides an overview of the status of Latin American biodiversity and ecosystems, drawing on indicators from national and international sources. It examines the main pressures on these ecosystems resulting from forestry and agriculture, mining, energy and infrastructure development, invasive species, desertification and climate change. The chapter includes an overview of the region's main biodiversity hotspots.

The statistical data for Israel are supplied by and under the responsibility of the relevant Israeli authorities. The use of such data by the OECD is without prejudice to the status of the Golan Heights, East Jerusalem and Israeli settlements in the West Bank under the terms of international law.

2.1. Status and trends

Latin America is one of the most important regions of the world in terms of biodiversity and ecosystems. Latin America and the Caribbean hold 40% of the world's biological diversity, eleven of the 14 terrestrial biomes, and the second largest reef system worldwide (IDB, 2015). The region holds more than 30% of global freshwater, 50% of tropical forests, 33% of mammals, 35% of reptilian species, 41% of birds and 50% of amphibians (UNEP, 2010). Six of the world's 17 "megadiverse" countries are found in Latin America – Mexico, Colombia, Venezuela, Ecuador, Peru and Brazil – selected based on the proportion of species that are endemic (found nowhere else in the world) and the presence of important marine ecosystems (Biodiversity A-Z, 2014). Biodiversity hotspots – characterised by high degrees of endemism and biodiversity loss – extend across many South American countries, including Paraguay, Uruguay, Chile, Argentina, Bolivia, Brazil, and all countries of Central America.

Box 2.1. Biodiversity hotspots of Latin America

North and Central America

- *Madrean Pine-Oak Woodlands*: Stretching across Mexico's main mountain chains and into the southern United States, this region holds one quarter of Mexico's plant species. The pine forests of Michoacán – threatened by excessive logging – provide a wintering site for the annual migration of millions of monarch butterflies.

- *Mesoamerica*: This region is the third largest hotspot in the world and spans Mexico and most of Central America. It is a corridor for many neotropical migrant bird species, has over 17 000 plant species and provides habitat for amphibians. Species are threatened by habitat loss, fungal disease and climate change.

- *North American Coastal Plain*: A newly announced hotspot in 2016 covers the south-eastern United States and north-eastern Mexico, and is characterised by more than 1 500 endemic vascular plants and 70% habitat loss. Population growth, sea-level rise and loss of historic dispersal corridors are threatening species.

South America

- *Atlantic Forest*: The Atlantic Forest region extends along Brazil's coast, inland to eastern Paraguay and into Argentina and Uruguay. Over 40% of the 20 000 plant species, and 15% of the 930 bird species, are endemic to the region, and the 8% of original forest remaining is threatened by agriculture and urban expansion.

- *Cerrado*: Covering 21% of Brazil, the Cerrado is the most extensive woodland-savannah in South America. It is home to species such as the giant anteater, giant armadillo, jaguar and maned wolf. Agriculture and ranching pose threats to biodiversity in the region.

    Chilean Winter Rainfall-Valdivian Forests: This hotspot in central Chile encompasses 40% of the country, divided between a Mediterranean-type climate

> and winter-rainfall deserts. Species such as the Araucaria tree, Andean cat, and endemic reptiles and amphibians are threatened by agriculture and urban development.
>
> - *Tropical Andes*: Stretching from Venezuela through Colombia, Ecuador, Peru and Bolivia into Chile and Argentina, this region is one of the richest and most diverse on Earth. It is home to a number of endemic plants, mammals and birds, and the largest variety of amphibians in the world, threatened from mining, oil, forestry, and plantations.
>
> - *Tumbes-Chocó-Magdalena*: Extending from the Panama Canal, into Colombia, Ecuador and Peru, this hotspot includes habitats such as mangroves, beaches, rocky shorelines, coastal wilderness, rain forests and South America's only remaining coastal dry forest. Threats include urbanisation, hunting and deforestation.
>
> *Source*: CEPF (2016a), *Biodiversity Hotspots*, www.cepf.net/resources/hotspots/Pages/default.aspx.

2.1.1. Forests

Forests cover nearly half of the Latin American continent, which is large in international comparison. Between 1990 and 2005, Latin America and the Caribbean lost nearly 69 million ha of forest, or 7% of the region's forest cover (UNEP, 2010). On average, forest area has declined by 0.4% per year in South America, compared to 0.1% globally (FAO, 2015). While the forest loss on the continent has slowed in recent years, deforestation rates remain among the highest in the world, constituting one of the greatest challenges to biodiversity conservation. The deceleration of forest loss was much driven by Brazil, which reduced deforestation in the Amazon from 27 700 km^2 in 2004 to 4 800 km^2 in 2014 (OECD, 2015). Bolivia, Colombia and Mexico have also slowed the rate of deforestation, while Chile, Costa Rica and Uruguay are expanding their forest area (Figure 2.1). Deforestation rates remain very high in much of Central America (FAO, 2015).

Figure 2.1. Forest loss remains high

Annual percentage change in forest area in selected Latin American countries

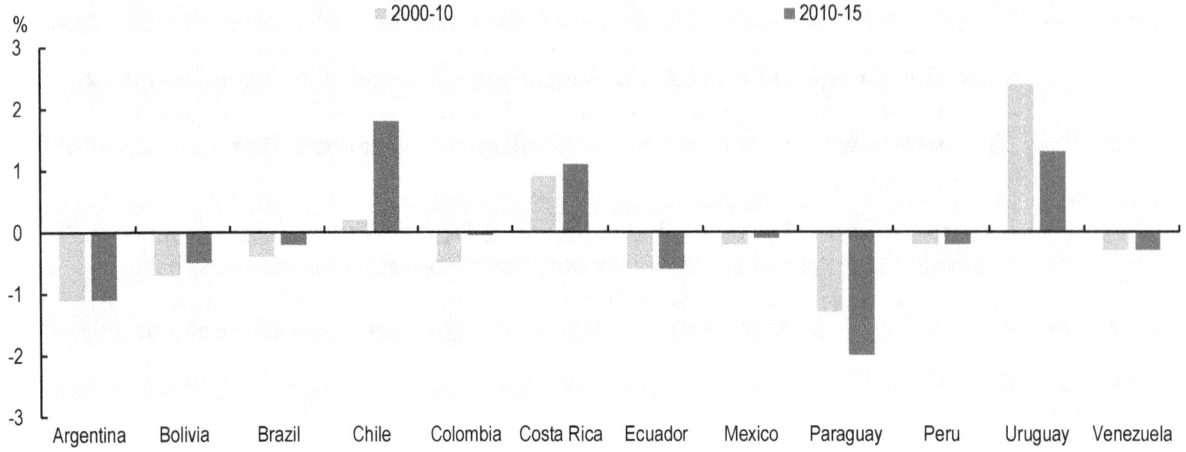

Source: FAO (2015), *Global Forest Resource Assessment 2015*, www.fao.org/forest-resources-assessment/en/

StatLink https://doi.org/10.1787/888933886018

2.1.2. Marine ecosystems

Latin American countries mainly rank in the middle of the 221 countries included in the 2017 Ocean Health Index assessment, which evaluates marine ecosystems around the world. Chile and Ecuador are among the region's leaders, ranking 70th and 82nd respectively, while Colombia and Venezuela are among the worst performers. Ocean Health Index scores – which include biodiversity, ecosystem and economic criteria – range between a low of 60 for Colombia and a high of 71 for Chile and Easter Island (Figure 2.2). Most Latin American countries are below the global score of 70. Data limitations continue to be a challenge in fully assessing some countries, however (Ocean Health Index, 2017).

Figure 2.2. Ocean Health Index scores are deteriorating

Ocean Health Index, selected Latin American countries

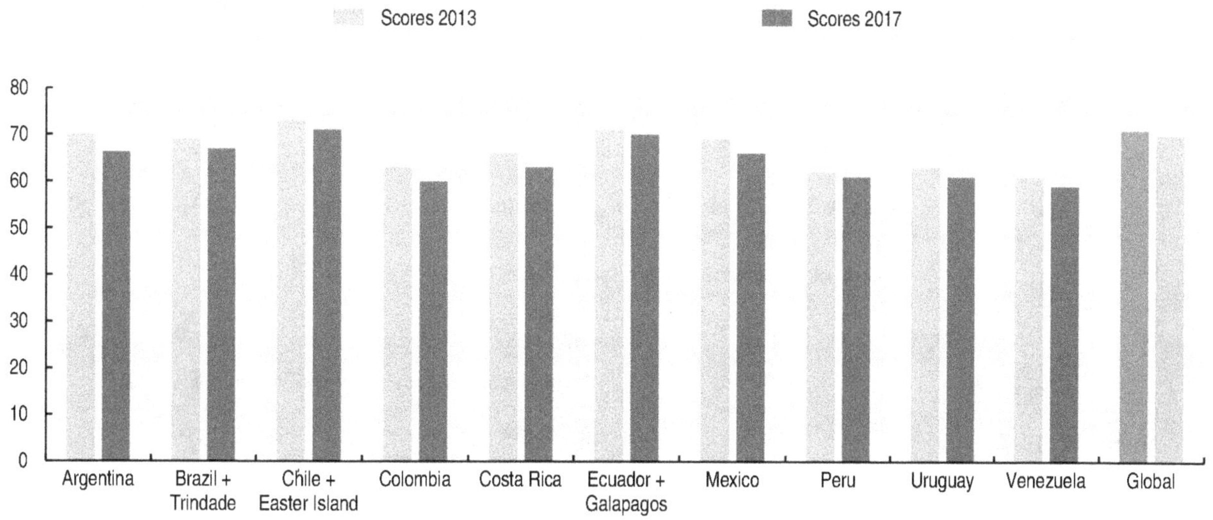

Note: Overall scores are based on several biodiversity, ecosystem and economic criteria, including biodiversity, clean waters, carbon storage, artisanal fishing opportunities and tourism and recreation. The biodiversity criterion measures how successfully the richness and variety of marine life is being maintained in the country. The overall scores are out of a maximum of 100.
Source: Ocean Health Index (2017), *Ocean Health Index 2017*, www.oceanhealthindex.org.

StatLink https://doi.org/10.1787/888933886037

2.1.3. Inland and aquatic ecosystems

Although Latin America holds more than 30% of the world's freshwater in its lakes, rivers, wetlands and aquifers, water resources are unequally distributed. Many arid and semi-arid regions are expected to face increasing challenges with water availability that will impact biodiversity, economic growth and drinking water supplies as water demand grows and climate change exacerbates water scarcity (UNEP, 2010).

The Andes Mountains in South America hold 90% of the world's tropical glaciers, which are a vital source of fresh water for humans and biodiversity in the sub-region. The Intergovernmental Panel on Climate Change predicts that most of the glaciers will melt by 2040 (UNEP, 2010). Northern and central Chile is facing growing water scarcity challenges due to climate change as well as water-intensive mining activities, agriculture and population growth. These are threatening wetlands and the birds, amphibians and other species dependent on them.

2.1.4. Species

Latin American countries have some of the highest numbers of threatened species in the world, and many more have not yet been assessed. Extinction risk is particularly high among coral, tree, and amphibian species (UNEP, 2010). Latin America also has high proportions of endemic species that are found nowhere else in the world. For example, 25% of the 31 000 described species in Chile are endemic.

The Red List of threatened species of the International Union for the Conservation of Nature (IUCN) shows that Ecuador, Mexico, Brazil, Colombia and Peru have the highest number of threatened species in the region (Figure 2.3). For example, in Brazil, the 2014 list of threatened flora species indicates that 46% of the 4 600 evaluated plant species are threatened under various risk categories (OECD, 2015). Ecuador and Costa Rica have some of the highest shares of threatened species compared to the total number of known endemic species in their country (Figure 2.3). Both countries see more than two-thirds of their endemic birds under threat, a larger share than any other OECD country (OECD, 2018). However, these numbers may not be reflective of the true status as many countries have only assessed a small portion of known species. Chile, for example, has only classified 3.5% of known species (see Section 3.6).

Figure 2.3. The number of threatened species is high

Number of threatened species in selected Latin American countries

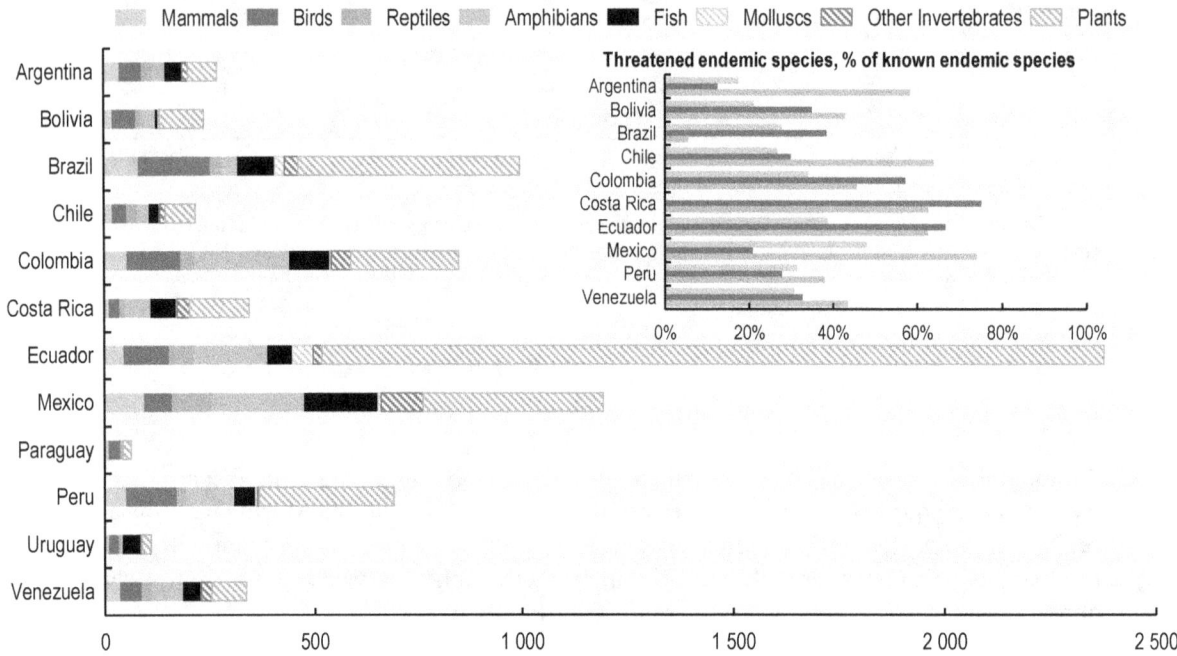

Note: The number of species identified as threatened is also a function of how many have been assessed in the country. The IUCN notes that there are many species that have not yet been assessed, particularly reptiles, fish, molluscs, other invertebrates and plants.
Source: IUCN (2018), *IUCN Red List of Threatened Species*, www.iucnredlist.org/.
StatLink https://doi.org/10.1787/888933886056

2.2. Key pressures

Pressures on biodiversity and ecosystem services are growing quickly in many regions of Latin America as a result of the scale and pace of economic and population growth. While significant progress has been made, in a number of cases biodiversity conservation and sustainable use policies have not evolved fast enough to prevent biodiversity loss and ecosystem degradation. Between 1990 and 2010, Latin America's population grew by more than 30%, and GDP in the region increased 87%. By 2030, the population is expected to reach 691 million (from 633 million in 2015), and GDP is expected to reach USD 9.2 trillion (from USD 6.2 trillion in 2015) (IDB, 2015; IDB, 2016). Forestry, agriculture, mining, and energy extraction and infrastructure are some of the key sectoral drivers of biodiversity loss. These are outlined below, and discussed in greater detail in Chapter 6 of the report.

2.2.1. Forestry and agriculture

Deforestation remains one of the greatest pressures on biodiversity in Latin America. This is predominantly driven by the desire to convert forest into agricultural land to grow commercial crops (e.g. soya, biofuels, fruits, vegetables, flowers) and raise livestock for export (UNEP, 2010). In the Cerrado region of Brazil (a biodiversity hotspot) large-scale land clearing for agriculture has left only around 20% of the original vegetation intact (CEPF, 2016b). Agricultural expansion also caused over 90% of deforestation in the Peruvian Amazon. Unclear or lack of land tenure, as well as illegal activities (logging, mining, illegal crops, wildlife traffic) are contributing to deforestation (OECD/ECLAC, 2017; OECD, 2015). In Colombia 40-50% of timber is harvested illegally (MADS, 2012). Illicit crop cultivation is a challenge in both Colombia and Peru. Forest fires are also a major source of forest loss, particularly in Chile and Brazil. Chile has an estimated 5 000 fires annually, causing about USD 50 million of financial loss per year (OECD/ECLAC, 2016). Fragmented and lost forest areas not only threaten the viability of a number of species, they can also have adverse impacts on the water quality of watersheds, lead to higher soil erosion and increase greenhouse gas (GHG) emissions. Agriculture itself is also a significant threat to biodiversity, as a result of overgrazing, pesticide and fertiliser use, and high water use.

2.2.2. Mining, energy and infrastructure development

Mining, oil and gas extraction, and electricity production have environmental impacts that represent significant risks to biodiversity such as high groundwater extraction, land-use change, soil and water contamination and hazardous waste generation (e.g. in tailings ponds from mining). In Chile, mining activity has led to elevated copper and salinity levels in some rivers. Expansion of pipeline infrastructure to transport oil and gas to markets can also lead to spills and disruption of ecosystems. For example, oil company PetroPeru experienced three oil spills between January 2016 and June 2016 in the Peruvian Amazon region. Hydroelectric development, which is significant in Latin America, can result in displacement of people and destruction of natural habitat for the creation of reservoirs. Expanding road infrastructure, driven by urban growth as well as by resource extraction and energy development, is further threatening biodiversity by creating access to previously remote areas, allowing others to clear land for subsistence agriculture or illegal logging (UNEP, 2010). This happened in Brazil and Colombia where deforestation has often occurred along new road as a consequence of easier access to the forest. In Colombia about 60% of roads are built by municipalities and departments, often with weak planning or technical design and therefore do not incorporate environmental considerations. This situation has been exacerbated after the end of armed conflict by the return of population to remote and areas.

2.2.3. Invasive species

Invasive Alien Species are a mounting threat to biodiversity. They can out-compete native species for space and resources, be predatory to native species, and/or introduce disease. They can also cause economic harm by damaging agricultural production, forestry, fishing and water supplies (ICSU, 2009). In Brazil, the presence of invasive species is estimated to cause an annual loss of USD 43 billion (OECD, 2015).

Many invasive species were introduced intentionally. Indeed, it is estimated that three quarters of invasive species found in Brazil were introduced deliberately, mainly for agriculture and ornamental use. The Canadian beaver was introduced to the island of Tierra del Fuego bordering Argentina and Chile in 1946 with the intention of fostering a fur trade. The beaver grew in population and now numbers in the tens of thousands, spreading to other islands and areas north of the Strait of Magellan. The beaver is particularly destructive in the area, because Patagonian forests do not grow back in the same way as North American trees. Beaver ponds are also causing rivers to retain more organic matter, altering the watershed's carbon cycle (OECD/ECLAC, 2016). Similarly, the expansion of the Giant African land snail, initially brought to Brazil for commercial purposes (for the development of an "escargot market") is now causing environmental damage in several countries across the continent.

Knowledge of invasive species is limited and uneven across Latin American countries. The Global Invasive Species Database developed by the IUCN Invasive Species Specialist Group provides a good comparable indication of invasive species present in Latin American countries, though it does not include all invasive species. According to the database, Latin America has 54 of the top 100 of the world's worst invaders, and a greater prevalence of invasive species in categories such as trees, vines, climbers, mammals, fish, amphibians, and insects (ICSU, 2009). Mexico, Brazil and Argentina showcase the largest numbers of invasive species (Figure 2.4), although this may also reflect greater data availability. In general, the numbers of invasive species listed on the Global Invasive Species Database are lower in Latin American countries than in OECD countries such as Australia (409), France (254) and Canada (243) (IISG-IUCN, 2016).

Figure 2.4. Invasive species are a threat to biodiversity

Number of invasive species in selected Latin American countries, 2016 or latest available year

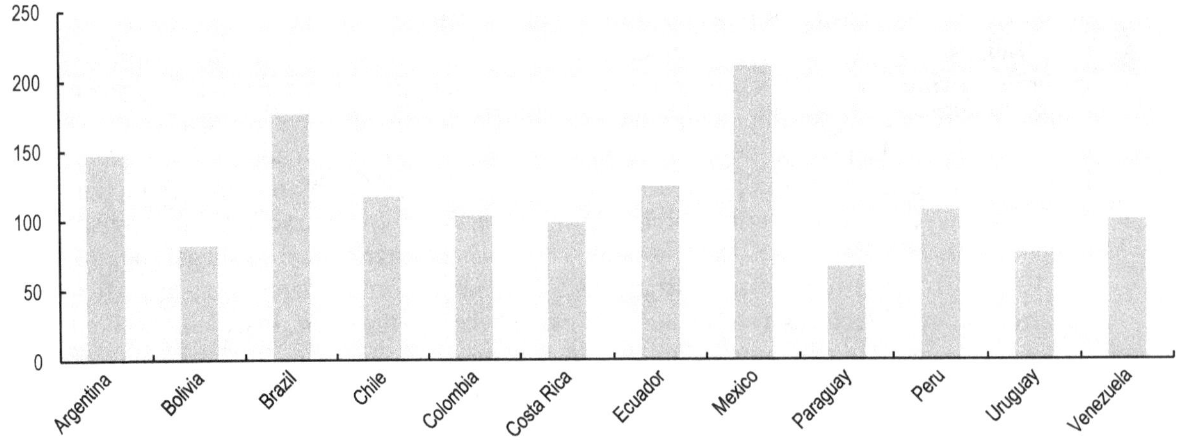

Source: ISSG-IUCN (2018), *Global Invasive Species Database*, http://www.iucngisd.org/gisd/.
StatLink https://doi.org/10.1787/888933886075

2.2.4. Desertification

Latin America is particularly vulnerable to desertification, with about one-quarter of the territory consisting of desert and drylands. Most of Mexico is arid and semi-arid. Southern Ecuador, the Peruvian shoreline and northern Chile have hyper-arid deserts. High and dry plains of the Andean mountains cover large areas of Peru, Bolivia, Chile and

Argentina. To the east of the Andes, an arid region reaches from Paraguay into Patagonia in southern Argentina. Northeast Brazil contains semi-arid zones with tropical savannahs (UNCCD, 2007). Land degradation, overgrazing, deforestation, forest fires, excessive water use for irrigation and droughts exacerbated by climate change make biodiversity and human populations in these regions extremely vulnerable. Up to 50% of agricultural land in Latin America is at risk of desertification by the 2050s (IPCC, 2007). This has strong socio-economic impacts. In Peru, for example, most areas where soil quality is deteriorating are inhabited by populations with medium to low development indices (OECD/ECLAC, 2017). Several strategic ecosystems in Colombia are threatened by desertification, with the Caribbean area being the most vulnerable.

2.2.5. Other

Fishing and aquaculture are important industries in Latin America, yet overfishing, bycatch, illegal fishing and pollution from aquaculture are placing substantial pressure on marine and coastal ecosystems. Untreated waste, urban and industrial wastewater effluent and unsustainable tourism are placing further pressures on these ecosystems. As a large percentage of Latin America's population and development activities, including the main transport nodes, are concentrated in coastal areas, coastal development is also a driver of biodiversity loss. Inland aquatic ecosystems are threatened by pollution stemming from agriculture and aquaculture, existing and abandoned mines, oil extraction and wastewater.

References

Biodiversity A-Z (2014), *Megadiverse Countries*, United Nations Environment Programme World Conservation Monitoring Centre, Cambridge, www.biodiversitya-z.org/content/megadiverse-countries.

CEPF (2016a), *Biodiversity Hotspots*, Critical Ecosystem Partnership Fund, Conservation International, Arlington, www.cepf.net/resources/hotspots/Pages/default.aspx.

CEPF (2016b), "Cerrado", Critical Ecosystem Partnership Fund, Conservation International, Arlington, www.cepf.net/resources/hotspots/South-America/Pages/Cerrado.aspx.

FAO (2015), *Global Forest Resources Assessment 2015*, Food and Agriculture Organisation, Rome, www.fao.org/forest-resources-assessment/en/.

ICSU (2009), *Biodiversity Knowledge, Scope of Research and Priority Areas: An Assessment for Latin America and the Caribbean*, International Council for Science: Regional Office for Latin America and the Caribbean, Mexico City, www.icsu.org/icsu-latin-america/publications/reports-and-reviews/biodiversity-knowledge/Final%20Report_biodiversity_final_completo.pdf.

IDB (2016), *Latin American and the Caribbean 2030: Future Scenarios*, Inter-American Development Bank and The Atlantic Council of the United States, Washington D.C., http://publications.atlanticcouncil.org/lac2030/wp-content/uploads/2016/12/LAC2030-Report-Final.pdf.

IDB (2015), *Biodiversity and Ecosystem Services Program: An Overview*, Inter-American Development Bank, Washington D.C., http://idbdocs.iadb.org/wsdocs/getdocument.aspx?docnum=38186826.

ISSG-IUCN (2018), *Global Invasive Species Database*, Invasive Species Specialist Group of the IUCN, www.iucngisd.org/gisd/.

IPCC (2007), Fourth Assessment Report of the Intergovernmental Panel on Climate Change, Core Writing Team, Pachauri, R.K. and Reisinger, A. (eds.) IPCC, Geneva, Switzerland.

MADS (2012), *Politica Nacional para la Gestion Integral de la Biodiversidad y Sus Servicios Ecosistemicos (National Policy for the Integral Management of Biodiversity and its Ecosystem Services)*, Ministry of Environment and Sustainable Development, Bogota, http://humboldt.org.co/images/pdf/PNGIBSE_espa%C3%B1ol_web.pdf.

MINAM (2014), *Informe Nacional del Estado del Ambiente 2012-2013 (National Report on the State of the Environment 2012-2013)*, Ministry of Environment of Peru, Lima.

Ocean Health Index (2017), *Ocean Health Index 2017*, www.oceanhealthindex.org/.

OECD (2018), "Biodiversity: Threatened species", *OECD Environment Statistics* (database), https://doi.org/10.1787/data-00605-en (accessed on 16 July 2018).

OECD (2015), *OECD Environmental Performance Reviews: Brazil 2015*, OECD Publishing, Paris, http://dx.doi.org/10.1787/9789264240094-en.

OECD (2013), *OECD Environmental Performance Reviews: Mexico 2013*, OECD Publishing, Paris, http://dx.doi.org/10.1787/9789264180109-en.

OECD/ECLAC (2017), *OECD Environmental Performance Reviews: Peru 2017*, OECD Publishing, Paris, http://dx.doi.org/10.1787/9789264283138-en.

OECD/ECLAC (2016), *OECD Environmental Performance Reviews: Chile 2016*, OECD Publishing, Paris, http://dx.doi.org/10.1787/9789264252615-en.

OECD/ECLAC (2014), *OECD Environmental Performance Reviews: Colombia 2014*, OECD Publishing, Paris, http://dx.doi.org/10.1787/9789264208292-en.

UNCCD (2007), Combating desertification in Latin America and the Caribbean: Fact Sheet, United Nations Convention to Combat Desertification, Bonn, www.unccd.int/Lists/SiteDocumentLibrary/Publications/Fact_sheet_13eng.pdf.

UNEP (2010), *State of Biodiversity in Latin America and the Caribbean*, United Nations Environment Programme, Panama City and Nairobi, www.unep.org/delc/Portals/119/LatinAmerica_StateofBiodiv.pdf.

3. Institutional and policy frameworks

This chapter examines progress in the governance of biodiversity in Latin America. It begins with an overview of the institutional settings in the reviewed countries, including mechanisms for stakeholder participation. The chapter then discusses the overarching biodiversity strategies, legislation, goals and targets. The role of regional and global biodiversity initiatives is also considered. The final section discusses the status of data and knowledge, including on the economic and social value of biodiversity.

The statistical data for Israel are supplied by and under the responsibility of the relevant Israeli authorities. The use of such data by the OECD is without prejudice to the status of the Golan Heights, East Jerusalem and Israeli settlements in the West Bank under the terms of international law.

3.1. Introduction

Institutional and policy frameworks for biodiversity conservation and sustainable use have improved significantly over the past decade. Environment Ministries are leading the development of new policies and programmes, co-ordination mechanisms are improving and dedicated agencies are increasingly being established to manage protected areas. Efforts to improve the participation of stakeholders, indigenous peoples and local communities in project and land-use decision making are also accelerating.

International agreements and organisations are helping to drive additional strategies and action plans. The UN Convention on Biological Diversity (CBD) in particular has encouraged countries to update existing strategies to incorporate the 2011-20 Aichi Targets. Regional agreements such as the Latin America Initiative for Sustainable Development (ILAC) have also been influential, helping to address shared ecosystems and improve information sharing and co-operation.

However, while strategies are proliferating, implementation has remained a challenge. Few countries have managed to comprehensively and effectively integrate biodiversity into sectoral policies. As in other parts of the world, actions are also stymied by lack of adequate financial and human resources, low political priority for biodiversity and lack of capacity and co-ordination across national and regional authorities. A lot of work also remains to rebuild trust with communities regarding decision-making processes, in order to reduce environmental conflict. There are also significant differences across countries in the amount and quality of biodiversity data available.

3.2. Governance and institutions

Effective governance of biodiversity policy development and implementation is essential to improving conservation and sustainable use in Latin America. Significant efforts are being made to strengthen the institutional frameworks for biodiversity. In all five of the EPR countries, the Ministry of Environment – recently established in several countries – is responsible for overall biodiversity policy development. Additionally, co-ordination across ministries has improved, with the establishment of committees and other bodies with broad membership focused on biodiversity.

Four of the five EPR countries have established dedicated agencies for implementation and management of protected areas. Dedicated agencies allow for greater co-ordination, efficiency and focus than fragmented approaches across multiple institutions. Brazil established the Chico Mendes Institute for Biodiversity Conservation (ICMBio) in 2007 to improve the management of an increasing number of federal protected areas and to separate it from the licensing, monitoring and enforcement of environmental legislation (previously both functions were performed by the federal environment agency). This helped increase transparency of the national protected areas system (OECD, 2015). Colombia has a National Parks Authority, Peru has a National Service for State-Protected Natural Areas (SERNANP) and Mexico has a National Commission of Natural Protected Areas (OECD/ECLAC, 2017, 2014; OECD, 2013). Chile has pending legislation to establish a Biodiversity and Protected Areas Service (SBAP) that will support more effective and efficient biodiversity governance, consolidating activities currently undertaken by multiple organisations, improving enforcement of protected area management plans, and monitoring and inventorying species and ecosystems (OECD/ECLAC, 2016).

Some countries have been experimenting with the decentralisation of environmental responsibilities, with mixed results. Peru transferred the bulk of environmental responsibilities, previously in the hands of sector authorities, to a newly created Ministry of Environment in 2008, while also transferring additional responsibilities to sub-national and local authorities (OECD/ECLAC, 2017). Colombia relies on Autonomous Regional Corporations for biodiversity protection at the local level (OECD/ECLAC, 2014). While there are significant practical advantages to decentralising responsibilities and providing a clear role for local authorities, the way this has been done in Colombia and Peru has led to uneven performance, inconsistent approaches and inadequate human and financial resources in some areas. This highlights the need for mechanisms fostering a better territorial balance across local constituencies by providing support to regional and local governments most in need of strengthening in their technical and financing capacities.

Inter-ministerial commissions are a common tool to ensure co-ordination and with sectoral ministries. Peru's National Commission on Biological Diversity (CONADIB), which consist of representatives from the public and private sector, monitors the implementation of the commitments arising from CBD and related agreements; it also serves as an advisory and co-ordination body on the sustainable use of biodiversity. Mexico has established a National Commission for the Knowledge and Use of Biodiversity (CONABIO), with representation from ten ministries, to improve co-ordination. However, the organisation is more of an applied research organisation than a policy formulation body. The establishment of such a body – potentially building on the model of Mexico's Inter-Ministerial Commission on Climate Change (Box 3.1) – could further improve co-ordination across institutions and facilitate effective integration of biodiversity into other sector policies. Colombia has established an Inter-sectorial Commission on Climate Change as well as commissions for deforestation control and the sustainable development goals, which are contributing to biodiversity protection. While co-ordination is improving, few countries have comprehensively and effectively integrated biodiversity into sectoral policies (Chapter 6).

Box 3.1. Mexico's Inter-Ministerial Commission on Climate Change

Mexico's Inter-Ministerial Commission on Climate Change (CICC) is responsible for formulating national policies and strategies to address climate change, and works to ensure that ministries with responsibilities in relation to greenhouse gas (GHG) emissions lead implementation. It is supported by working groups on seven different policy areas, and consultative advisory bodies to engage experts and ensure societal participation. This institutional framework, combined with financial resources for organisations involved, has driven the advancement of climate change policy development and implementation.

Based on the success of the CICC, the EPR of Mexico suggested that an Inter-Ministerial Commission on Biodiversity responsible for formulating new policies and strategies could be created and linked to the existing National Commission for Knowledge and Use of Biodiversity (CONABIO), an applied research organisation that sponsors basic research, compiles and disseminates information and develops capacity.

Source: OECD (2013), *OECD Environmental Performance Reviews: Mexico 2013*; CONABIO (2016), *About Us*, www.conabio.gob.mx/web/conocenos/quienes_somos_ingles.html.

3.3. Stakeholder participation and engagement of indigenous peoples and traditional communities

Successfully addressing threats to biodiversity, and developing cost-effective and long-term approaches to conservation and sustainable use, will increasingly require the involvement of the private sector, indigenous peoples and local communities, and non-governmental organisations (NGOs) (UNEP, 2012). Approaches that leverage private sector financial resources, align environmental and economic objectives, empower and provide opportunities for indigenous peoples and local communities and engage NGOs help extend biodiversity policy beyond government-led, isolated policies and remote protected areas towards an integrated, more effective strategy that reduces conflict and supports positive environmental, economic and social outcomes. The Aichi Biodiversity Targets under the CBD Strategic Plan for Biodiversity 2011-20 include stakeholder involvement and highlight the importance of including indigenous and local communities in planning and implementation (UNEP, 2012).

Several Latin American countries have incorporated Principle 10 of the 2012 Rio Declaration on Environment and Development into their domestic legislation. The Principle encourages measures at the regional, national, sub-national and local levels to promote access to environmental information, promote public participation in decision-making, and ensure access to justice. As a result, there is a greater prevalence of consultations, public hearings and NGO inclusion in environmental management councils, as well as more environmental courts and tribunals to address cases of environmental conflict. For example, Mexico has 14 consultative bodies to facilitate public participation in environmental matters at the national level, focussing on themes including wildlife management and conservation, natural protected areas, forestry, climate change and water management (OECD, 2013). Colombia recently created the Inter-sectorial Table for Environmental Democracy for the fulfilment of the Escazú Agreement (a regional legally binding agreement that implements Principle 10). The agreement, which was signed by 24 Latin America and the Caribbean countries in March 2018, is also the first agreement worldwide to include dispositions on the protection of defenders of human rights in environmental matters. Implementation challenges remain, however, given the complexity of societies and certain land-use and resource conflicts (UNEP, 2016).

Brazil's 2007 National Policy for the Sustainable Development of Traditional Peoples and Communities and the 2012 National Policy on Territorial and Environmental Management of Indigenous Lands promote the sustainable use of natural resources on indigenous lands and defend the traditional knowledge of indigenous communities. The policies have helped improve relationships between environmental NGOs, the government and organisations working with indigenous peoples, though conflicts over land use rights can still arise with loggers, farmers and miners (OECD, 2015). Roughly 13% of Brazil's territory is protected by the designation of about 600 indigenous lands, most of which are located in the Amazon. The lands are considered protected areas under the CBD because of the long-standing tradition of indigenous communities to sustainably use natural resources. Deforestation rates on indigenous lands are the lowest in the country. Colombia recently designated ancient territories in the Sierra Nevada of Santa Marta as traditional area of spiritual, cultural and environmental protection with the aim to defend the traditional knowledge of indigenous communities and promote the sustainable use of natural resources on indigenous lands.

3.4. Biodiversity strategies and legislation

Latin American countries have made significant progress over the last 15 years in developing overarching biodiversity strategies, legislation, goals and targets. Brazil, Chile, Colombia, Mexico and Peru have all updated their national biodiversity strategies with a view to incorporate international commitments such as the 2011-20 Aichi Targets under the CBD. However, successful implementation of action plans continues to be a challenge, as a result of a lack of adequate financial and human resources, low political priority assigned to biodiversity and a lack of capacity and effective co-ordination across ministries and regions.

Brazil's biodiversity strategies have shifted from a "fence-and-protect" approach to one that favours the sustainable use of biological resources and identifies biodiversity priority regions and recognises the role of rural, traditional and indigenous communities in preserving ecosystems. In 2013, Brazil developed five strategic objectives and 20 national biodiversity targets closely aligned with the CBD Strategic Plan 2011-20, which was based on a broad consultation process. A multi-stakeholder panel (PainelBio) is leading a process to define indicators to monitor progress. Biodiversity efforts are also supported by a comprehensive legislative framework, including the 2000 law establishing the National System of Protected Areas (SNUC) and the 2012 Forest Code that regulates the protection of forests on private properties (OECD, 2015).

Chile updated its National Biodiversity Strategy, first published in 2003, in early 2018. The new strategy (which covers the period 2017 to 2030) incorporates the Aichi Targets and corrects some of the implementation challenges that arose in the first strategy. The new strategy shifts the focus from direct actions to enablers such as knowledge, capacity, awareness and education along with clear identification of financial requirements. It also increases the emphasis on ecosystem restoration and connectivity (OECD/ECLAC, 2016). Chile also has several policies, strategies and plans dealing with specific biodiversity-related issues, such as a national policy for the protection of threatened species (2005) and the National Glacier Strategy and Policy (2009).

Mexico's first National Biodiversity Strategy, developed in 2000, set out a 50-year vision to avert large-scale conversion of natural ecosystems. It focused on four areas: i) knowledge management; ii) valuation of biodiversity; iii) conservation; and iv) diversification of use (OECD, 2013). Mexico updated the strategy in 2016 and broadened its coverage so that its foundations are knowledge and education, communication and environmental culture, its pillars are conservation and restoration, sustainable use and management and attention to pressure factors, and its overarching roof is mainstreaming and governance. The strategy is accompanied by a plan for implementation, with 24 lines of action and 160 actions (Government of Mexico, 2016). Mexico also has a Strategic Forest Programme, which establishes targets up to 2025 to strengthen sustainable development of natural resources in forest ecosystems. It aims to establish plantations over a total area of 875 000 ha by 2025 and ensure that one-third of Mexico's territory is subject to some form of conservation and sustainable use regime (OECD, 2013).

Colombia has integrated biodiversity and sustainable development into its Constitution and into its National Development Plan. In 2012, the government also adopted a National Policy for the Integral Management of Biodiversity and Ecosystem Services (PNGIBSE) that seeks to influence environmental management in the country and updates previous policies to align them with CBD objectives and the 2011-20 Aichi Biodiversity Targets (OECD/ECLAC, 2014). Colombia's National Policy on Climate Change includes a strategic line for the management and conservation of ecosystems and their adaptation and mitigation services. In 2018, these policies where reinforced through the Moorlands and the Climate Change Laws.

Peru tasked its National Commission on Biological Diversity (CONADIB) with designing, updating and implementing its National Strategy on Biological Diversity, which runs until 2021. The Strategy also includes a 2014-18 Action Plan, which is supported by the Law on Conservation and Sustainable Use of Biological Diversity and its Regulation (OECD/ECLAC, 2017).

Given that countries can contain a wide range of ecosystems and environmental conditions, some countries have also developed biodiversity strategies at sub-national level. For example, Chile has 15 Regional Biodiversity Strategies that are currently being updated, and Mexico is developing state biodiversity strategies (OECD/ECLAC, 2016; OECD, 2013).

3.5. International and regional co-operation

Biodiversity policy in Latin American countries is significantly influenced by international and regional agreements and processes. International, regional, sub-regional and bilateral organisations and agreements offer significant potential for addressing pressures facing biodiversity in Latin America, and sharing information and best practices that can improve policy design and implementation. The CBD in particular has guided the strategies and commitments of signatory countries and helped to drive further domestic action, along the conventions on the strategic ecosystems of wetlands, wild species trade, ecosystems of forests, and desertification.

Most Latin American countries became parties to the 1992 CBD in the mid-1990s (Brazil was the first CBD signatory in 1992). Many are also parties to the 2003 Cartagena Protocol on Biosafety. Fewer Latin American countries have, however, ratified the 2014 Nagoya Protocol on Access to Genetic Resources and the Fair and Equitable Sharing of Benefits Arising from their Utilization (Nagoya Protocol on Access and Benefit Sharing). The Protocol is intended to help create incentives to conserve biodiversity, sustainably use its components and further enhance the contribution of biodiversity to sustainable development and well-being. Mexico, Peru and Colombia have ratified the protocol and begun to implement it domestically, while Brazil has signed the Protocol but not yet ratified it. However, Brazil has a national law on access and benefit sharing. Colombia has also undertaken initiatives to promote access and benefit sharing (Box 3.2). The Sustainable Development Goals (SDGs) and follow-up to the Rio+20 meeting in 2012 have also influenced approaches in several countries.

There are many initiatives at the regional level supporting information sharing and policy co-ordination and harmonisation related to biodiversity, such as the Latin American Initiative for Sustainable Development (ILAC), adopted in 2002 by the Forum of Ministers of the Environment of Latin America and the Caribbean. Sub-regional agreements such as the Amazon Cooperation Treaty, the Andean Community, the Central American Commission for Environment and Development and the Meso-American Strategy for Environmental Sustainability as well as numerous river basin agreements and mechanisms, are helping to drive co-ordinated action to improve biodiversity conservation and sustainable use (UNEP, 2013). Bilateral agreements can also be an important mechanism for boosting capacity and sharing best practices. Chile, for example, has initiatives in place with Canada and the United States focused on improving the management of certain protected areas (OECD/ECLAC, 2016). Brazil is a party to 233 bilateral and multilateral co-operation agreements, of which 22% have environmental themes (OECD, 2015).

Box 3.2. Colombia's approach to access and benefit sharing

Equitable benefit sharing from the use of genetic resources is an issue in Colombia, as 27% of the country's area under protection is on indigenous reservations or collective territories. In addition, innovation in biotechnology is an engine of growth in development plans.

Between 2004 and 2011, Colombia signed 45 agreements on access to genetic resources for research purposes. In 2011, the government released a national strategy on biotechnology and sustainable use that aims to improve institutional capacity for commercial development of biotechnology from biodiversity, adopt a set of economic instruments to attract public investment and private companies interested in developing products, adapt and revise a regulatory framework for access to genetic resources, and evaluate the creation of a national bio-prospecting company.

Colombia established free, prior and informed consent for indigenous groups in law through the ratification of 1989 Indigenous and Tribal Peoples Convention (the International Labour Organisation Convention 169). The provision of information to indigenous groups and the right of ethnic groups to exploit resources by traditional methods are also recognised by law. However, the EPR of Colombia noted that experience with free, prior and informed consent in relation to extractive industries was mixed, and suggested a strengthening of the arrangements for enforcement of fair access. This would ensure that companies comply with requirements, and that local and ethnic groups retain access to areas they have traditionally used. At the same time, the increasing investment, commercialisation and involvement of the private sector in the use of genetic resources underlines the importance of adequate provision for fair and equitable benefit sharing. The EPR recommended a formal system of benefit sharing to be established.

Source: OECD/ECLAC (2014), *OECD Environmental Performance Reviews: Colombia 2014*.

3.6. Status of data and knowledge

Comprehensive and accurate data and knowledge of the status of ecosystems and species, expanded monitoring and reporting of trends, and better insight into the economic and social importance of biodiversity are essential to informing decision-making, building public consensus around biodiversity conservation and sustainable use, identifying priorities for action and effectively designing and implementing biodiversity policy.

Despite significant improvement in the extent and depth of environmental indicators in Latin America over the past decade, a lack of biodiversity knowledge remains a key challenge. Brazil is estimated to host nearly 44 000 plant species and more than 104 500 vertebrate and invertebrate species, yet as of 2014 only 12 000 fauna species had been assessed. As mentioned above, Chile has assessed about 1 000 species, or 3.5% of known species in the country. However, the National Institute for Amazon Research in Brazil – one of the world's largest and most important research institutions on tropical biology – is actively working to improve species inventories and disseminate scientific knowledge of the Amazon biome, and Chile's environment ministry has announced plans to move forward with a National Ecosystem Assessment in 2016 or 2017 so as to improve the knowledge base (OECD/ECLAC, 2016).

Knowledge of the status and trends in marine and freshwater ecosystems is particularly limited. For example, Chile has classified less than 4% of fish species, and a lack of continuous and comprehensive data on the status of water bodies and coastal areas is a serious obstacle to effective management of water resources in the country (OECD/ECLAC, 2016). Given the paucity of data on marine ecosystems, other Latin American countries would benefit from following in the footsteps of Colombia to conduct independent assessments of their marine ecosystems using the international Ocean Health Index methodology. Independent assessments use the same framework as the global assessments, but allow for exploration of variables influencing ocean health at the smaller scales where policy and management decisions are made (Ocean Health Index, 2015). This would help countries to understand where to focus protection efforts.

Mexico has one of the most developed systems of biodiversity information in Latin America. Its National Biodiversity Information System includes satellite imaging data, electronic cartography, data on species and an early warning fire detection system, with priority areas such as mangroves and cloud forests being the focus of ecosystem monitoring. There is also a National Forest Information System, which includes a forestry and soil inventory, and fishery data. Mexico's System of Information, Monitoring and Evaluation of Conservation is used to analyse the effectiveness and impact of public policy implementation in priority regions for conservation (OECD, 2013). Brazil is a world leader in satellite-based deforestation monitoring systems, providing an example of how technology can improve knowledge for effective decision making (Box 3.3). Colombia has a Monitoring System of Forest and Carbon which tracks changes in the coverage of natural forest as well as five national environmental research institutions, including one specialised on biodiversity (the Biological Resources Alexander von Humboldt Research Institute). Colombia has also made significant progress in the generation of information on wetlands. However, the country lacks a long-term research agenda.

Box 3.3. Brazil's deforestation monitoring systems

The National Institute for Space Research (INPE) has monitored forest cover in the Amazon region annually since 1988. This monitoring system was improved in 2002 with the adoption of digital classification of satellite images using the Amazon Programme on Deforestation Monitoring (PRODES) methodology. This new approach drastically improved the precision of deforestation monitoring. INPE also runs the Real Time Detection Programme (DETER), a deforestation monitoring system in the Amazon, which shows alerts every two to three days and has been a key support to strategic law enforcement actions. In addition, the DEGRAD system monitors forest degradation and the TerraClass analysis assesses land-use change in previously deforested areas (MMA, 2015). According to TerraClass data, about one-third of the Amazon cleared forest land has been recovering.

In addition to annual monitoring of the Amazon forest cover, in 2008 the Brazilian Institute of Environment and Renewable Natural Resources (IBAMA) started a satellite monitoring programme (Programme on Satellite Monitoring of Deforestation in Brazilian Biomes, or PMDBBS) for the other five terrestrial biomes. However, PRODES is more precise than the systems used by PMDBBS, and the data is not fully compatible. Therefore, INPE and IBAMA are collaborating to develop a monitoring system for the entire national territory to generate continuous and compatible data series on deforestation, vegetation cover and land use for all biomes.

Source: OECD (2015), *OECD Environmental Performance Reviews: Brazil 2015*.

Many of the benefits associated with biodiversity are not reflected in market prices. Economic valuation studies, which estimate the monetary value of the ecosystem services provided by biodiversity, can illustrate the importance of conservation and sustainable use while supporting better policy decisions. Mexico, Brazil and Chile have done several studies on the economic valuation of biodiversity (Box 3.4), but these are not yet used frequently in decision-making processes. Several Latin American countries are involved with the World Bank WAVES (Wealth Accounting and the Valuation of Ecosystem Services) project that aims to mainstream natural resources in development planning and national economic accounts (WAVES, 2016a). Colombia is one of the core implementing partners that has begun to put in place natural capital accounting, both to support biodiversity management and promote the sustainable use of biodiversity as an engine for development. It initially focused on three pilot watersheds before expanding to integrated national-level accounts for water, forests and land (WAVES, 2016b). Brazil launched a Natural Capital Initiative in 2013 and has made progress on including the value of water resources in national accounting and work is continuing on forest accounting (OECD, 2015). These experiences should be built upon to fully integrate the value of biodiversity and ecosystem services into national accounts.

Box 3.4. Economic valuation of biodiversity

In **Mexico**, protected areas provide an estimated USD 3.4 billion in economic benefits and cost savings as a result of storing carbon, protecting water supplies and supporting tourism. Every Mexican peso invested in protected areas generates 52 pesos to the economy. In Pacific mangrove areas, the value of ecosystem services is low (USD 1 per hectare), but could be as high as USD 77 per hectare if overexploitation of the fishery is addressed. The Mexican government has used economic valuation of biodiversity to inform the design of its Payment for Ecosystem Services programme and the level of access fees for protected areas.

In **Chile**, a study estimating the economic values of the Valdivian rainforest ecoregion found values of USD 3 742 per hectare for sustainable forest management, and USD 4 546 for old growth forests. The annual value of maintaining soil fertility was USD 26 per hectare. A 2010 study estimated the monetary value of ecosystem goods and services from Chile's National System of Protected Areas to be USD 2 million when considering formal protected areas, private conservation areas and priority sites for conservation. This value includes regulating services such as water purification and regulation, pollination, waste treatment, climate regulation, erosion control, species shelter and habitat. It captured direct uses such as supply of food and fibre, water, fuel, tourism and recreation and included the provision of genetic resources and cultural services.

Brazil's protected areas system is estimated to have prevented the release of about 2.8 billion tonnes of CO_2 into the atmosphere, which in monetary terms would correspond to BRL 96 billion. The economic gains from tourism in national parks is estimated at BRL 1.6 billion per year, and sustainable timber logging in the Amazon protected areas generates between BRL 1.2 billion and BRL 2.2 billion annually.

Source: OECD (2013), *OECD Environmental Performance Reviews: Mexico 2013*; Bezaury Creel, J.E. and L. Pabón Zamora (2009), *Valuation of Environmental Goods and Services Provided by Mexico's Protected Areas*; Nahuelhual L. et al. (2007), "Valuing ecosystem services of Chilean temperate rainforests", *Environment, Development and Sustainability*, Vol. 9/4, Springer, pp. 481-499 ; Medeiros, R. and C. Young (2011), *Contribuição das unidades de conservação brasileiras para a economia nacional: Relatório Final*; OECD (2015), *OECD Environmental Performance Reviews: Brazil 2015*; OECD/ECLAC (2016), *OECD Environmental Performance Reviews: Chile 2016*.

References

CONABIO (2016), *About Us*, National Commission for Knowledge and Use of Biodiversity, Mexico City, www.conabio.gob.mx/web/conocenos/quienes_somos_ingles.html.

Government of Mexico (2016), *Estratégia Nacional sobre Biodiversidad de México y Plan de Acción 2016-2030,* National Commission for Knowledge and Use of Biodiversity, México D.F., www.cbd.int/doc/world/mx/mx-nbsap-v2-es.pdf.

MMA (2015), Fifth National Report to the Convention on Biological Diversity, Ministry of the Environment, Brasília, www.cbd.int/doc/world/br/br-nr-05-en.pdf.

Ocean Health Index (2017), *Ocean Health Index 2017*, www.oceanhealthindex.org/.

OECD (2015), *OECD Environmental Performance Reviews: Brazil 2015*, OECD Publishing, Paris, http://dx.doi.org/10.1787/9789264240094-en.

OECD (2013), *OECD Environmental Performance Reviews: Mexico 2013*, OECD Publishing, Paris, http://dx.doi.org/10.1787/9789264180109-en.

OECD/ECLAC (2017), *OECD Environmental Performance Reviews: Peru 2017*, OECD Publishing, Paris, http://dx.doi.org/10.1787/9789264283138-en.

OECD/ECLAC (2016), *OECD Environmental Performance Reviews: Chile 2016*, OECD Publishing, Paris, http://dx.doi.org/10.1787/9789264252615-en.

OECD/ECLAC (2014), *OECD Environmental Performance Reviews: Colombia 2014*, OECD Publishing, Paris, http://dx.doi.org/10.1787/9789264208292-en.

UNEP (2016), Summary of the progress in the implementation of the decisions of the Nineteenth Meeting of the Forum of Ministers of Environment of Latin America and the Caribbean, UNEP/LAC-IGWG.XX/3, Twentieth Meeting of the Forum of Ministers of the Environment of Latin America and the Caribbean, www.pnuma.org/forodeministros/20-colombia/documentos/Summary_Report_on_Decisions_FINAL.pdf.

UNEP (2013), Review of existing intergovernmental priorities on sustainable development, with an emphasis on environment, in Latin America and the Caribbean, UNEP/LAC-IGWG.XIX/8, Nineteenth Meeting of the Forum of Ministers of Environment for Latin America and the Caribbean, www.pnuma.org/forodeministros/20-colombia/documentos/Regional_and_subregional_priorities_23_August_2013_3_.pdf.

UNEP (2012), *Global Environment Outlook 5: Chapter 12 Latin America and the Caribbean*, United Nations Environment Programme, Nairobi, http://web.unep.org/geo/sites/unep.org.geo/files/documents/geo5_report_c12.pdf.

WAVES (2016a), *Wealth Accounting and the Valuation of Ecosystem Services*, World Bank, Washington D.C., www.wavespartnership.org.

WAVES Partnership (2016b), *Colombia*, World Bank, Washington D.C., www.wavespartnership.org/colombia.

4. Policy instruments

This chapter provides an overview of the main policy instruments used for biodiversity conservation and sustainable use. It begins with protected areas – the most prominent instrument for biodiversity conservation in Latin America – followed by a discussion on other regulatory approaches such environmental impact assessments, strategic environmental assessments, land-use planning and zoning. The chapter then examines the use of economic instruments such as payments for ecosystem services, biodiversity offsets, tradable resource extraction quotas and fiscal incentives. The role of environmentally harmful subsidies is also reviewed. The final section discusses voluntary and information instruments, such as certification, eco-labelling and voluntary agreements.

The statistical data for Israel are supplied by and under the responsibility of the relevant Israeli authorities. The use of such data by the OECD is without prejudice to the status of the Golan Heights, East Jerusalem and Israeli settlements in the West Bank under the terms of international law.

4.1. Introduction

Latin American countries have made significant progress in putting in place policy instruments for biodiversity conservation and sustainable use over the past decade. To date, most countries have relied heavily on regulatory approaches, but they are beginning to implement more economic instruments as well as information and voluntary approaches (Table 4.1). Implementation, enforcement, monitoring, capacity and resourcing remain ongoing challenges.

Table 4.1. Policy instruments for biodiversity in Latin America

Regulatory instruments	Economic instruments	Voluntary, procurement and information approaches
Restrictions or prohibitions on use or on access	**Price-based instruments**	**Certifications**
Protected areas	Water abstraction and pollution charges	Forestry certification
Restrictions on trade in animal and wild plant specimens	Wastewater charges and fees	Sustainable wine certification
Set-aside native vegetation areas	Protected area entrance and concession fees	Organic farming
Regulation on access to genetic resources and benefit sharing	Subsidies for conservation practices (e.g. good forestry and agricultural practices)	Best aquaculture practices
Embargos on illegal deforestation	Removing environmentally-harmful subsidies	Green certification for coffee
Fishing restrictions	Payment for ecosystem services programmes	Eco-tourism certification
Water quality and emission standards	Biodiversity offsets or biobanks	
	Fishery buybacks	**Reporting/inventorying**
Planning and licensing instruments		Peat extraction
Zoning and land-use planning	**Market-based instruments**	Abandoned mines
Environmental impact assessment	Tradable development rights	Wetlands
Strategic environmental assessment	Markets for water use rights	
	Tradable Fishing quotas	**Voluntary Agreements, e.g.**
Permits		Clean production agreements (Chile)
Concessions for sustainable logging		Soya Moratorium (Brazil)
Fishing, hunting, logging permits		
		Green Public Procurement, e.g.
		National Plan to Promote Production Chain of Socio-Biodiversity Products (Brazil)

Source: adapted from Karousakis, K., et al. (2012), "Biodiversity", in *OECD Environmental Outlook to 2050: The Consequences of Inaction.*

Establishing new protected areas is one of the primary tools used. While the proportion of protected areas in Latin America is impressive, many countries are struggling to ensure the representativeness of all ecosystems and adequately resource effective management. Mandatory Environmental Impact Assessments (EIAs) remain one of the key tools available to mitigate the biodiversity impacts of major energy, mining, industrial and infrastructure projects, and Strategic Environmental Assessments (SEAs) have been used in several sectoral policies and land-use plans.

The usage of economic instruments is also growing in Latin America, with countries such as Mexico leading the use of Payment for Ecosystem Services (PES) systems that pay individuals or communities for conservation measures, and biodiversity offset regimes that undertake conservation actions to compensate for residual biodiversity loss from development sites. A number of Latin American countries also use water charges, water markets, forestry fees and tradable fishing and forestry quotas. Subsidies that provide incentives to promote sustainable use of biodiversity and ecosystems remain an important component of biodiversity policy in Latin America, particularly for rural and poor populations. However, it is also important to reform environmentally-harmful subsidies established for other purposes such as tax exemptions for fertilisers and pesticides and subsidies for irrigation infrastructure and small-scale mining.

Voluntary and information initiatives can also be important avenues to biodiversity conservation and sustainable use. Eco-labelling is becoming increasingly popular, particularly as export markets for forest, agriculture and aquaculture products demand more sustainable production methods.

4.2. Regulatory instruments

4.2.1. Protected areas

The number and size of protected areas has been increasing in Latin America over the past decade, and Central and South America now have the largest percentage of terrestrial protection in the world. In 2014, Central America had 28.2% of its terrestrial areas protected and South America had 25.0% protected. Comparable percentages for Europe, North America and Asia were only 13.6%, 14.4% and 12.4% respectively (Juffe-Bignoli et al., 2014). OECD countries together had 15.4% of their land protected in 2017 (OECD, 2018). Several countries in the region exceed the CBD Aichi Target of conserving 17% of terrestrial area and inland water by 2020 (Figure 4.1).

However, Central and South America are behind other regions in the creation of marine protected areas, and remain below the CBD Aichi Target to conserve 10% of coastal and marine areas by 2020. This can be attributed to the region's historical policy focus on slowing deforestation, a gap in policy responsibility between environment and fisheries ministries, a lack of data and knowledge to assess biodiversity priorities, and a lack of financial and human resources. In 2014, marine protected areas comprised 2.1% and 3.9% of total marine areas in Central and South America respectively, versus 3.9% in Europe, 6.9% in North America, and 4.5% in Asia (Juffe-Bignoli et al., 2014). The recent designation of large marine protected areas in Chile, Mexico and Brazil is, however, likely to improve South America's ranking (Figure 4.1).

Figure 4.1. Several countries exceed the CBD target for terrestrial areas

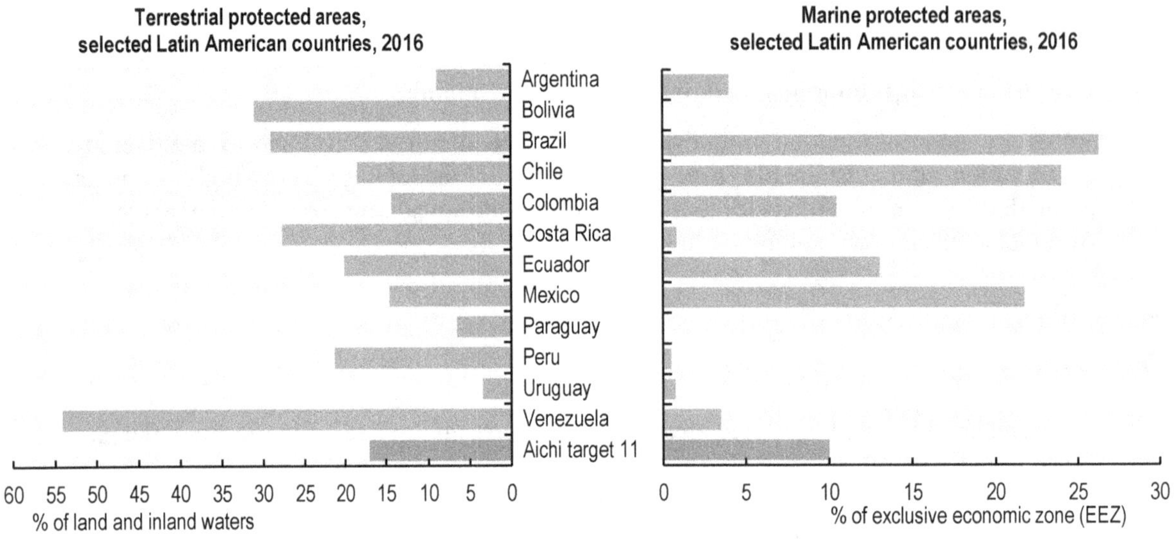

Terrestrial protected areas, selected Latin American countries, 2016 / Marine protected areas, selected Latin American countries, 2016

Countries listed: Argentina, Bolivia, Brazil, Chile, Colombia, Costa Rica, Ecuador, Mexico, Paraguay, Peru, Uruguay, Venezuela, Aichi target 11

% of land and inland waters / % of exclusive economic zone (EEZ)

Note: Data for Chile include the largest marine reserve in the Americas (Nazca-Desventuradas). Data for Brazil include two large mosaics of marine protected areas designated in March 2018 (Archipelago of Trindade and Martim Vaz and Monte Columbia and Archipelago of São Pedro and São Paulo).
Source: UNEP-WCMC and IUCN (2018), *The World Database on Protected Areas* (WDPA), January 2018. Available at: www.protectedplanet.net.

StatLink https://doi.org/10.1787/888933886094

Brazil has made one of the largest contributions to increase the global land area under protection since the turn of the century. Between 2000 and 2014, the number and extension of terrestrial protected areas in the country doubled, to reach a surface of almost three times the size of France. The Amazon Region Protected Areas (ARPA) programme has been at the heart of this progress (Box 4.1). About two thirds of the area under protection falls into the "sustainable use areas" category, which permits human settlements and various sustainable uses of natural resources. Allowing carefully controlled sustainable use of biodiversity in protected areas has proved to be helpful in overcoming political and social barriers to protected area expansion, partly as it is more compatible with traditional communities' rights. All protected areas are managed within the National System of Protected Areas (SNUC) which was established in 2000 to consolidate the pre-existing highly fragmented assortment of federal, state, municipal and private protected areas into one consistent framework (OECD, 2015).

Brazil also made a step increase with respect to marine conservation. In March 2018, the President signed decrees to create two large mosaics of marine protected areas: one for the Archipelago of Trindade and Martim Vaz and Monte Columbia situated in the Brazilian Exclusive Economic Zone (EEZ) of the coast of the State of Espírito Santo (with a total protection of 47.2 million ha); and one for the São Pedro and São Paulo Archipelago located in the extreme northeast of the EEZ, on the northeast coast of the State of Pernambuco (with a total protection of 44.9 million ha). The new areas lift the share of EEZ under protection from 1.6% to more than 26%. The government considers the establishment of these areas an important progress from both an environmental an economic perspective, as the areas will help contain the collapse of fish stocks. Both mosaics are a result of common efforts by the ministries of environment and defence (the limits of some areas are coincide with the limits of the EEZ).

> **Box 4.1. Amazon Region Protected Areas programme**
>
> The Amazon Region Protected Areas (ARPA) programme, launched in 2002, is one of the largest tropical forest conservation programmes in the world. It was created with the goal of expanding and strengthening the protected area system in the Amazon biome, including along the so-called "deforestation arc" and in areas expecting road infrastructure development. The ARPA programme made a significant contribution in fighting deforestation in the area.
>
> The programme had four major components: establishment of new areas; management and consolidation; financial sustainability; and co-ordination, management and monitoring. By 2015, the programme created more than 500 000 km² of protected areas in the biome. It has effectively supported the operation of protected areas by investing in basic infrastructure and capacity building. The ARPA programme has attracted substantial international finance for protected areas. However, the government envisions to shift funding from donation based to government financed over 25 years and has set up a transition fund for the purpose.
>
> *Source*: OECD (2015), *OECD Environmental Performance Reviews: Brazil 2015*.

A common challenge for Latin American countries is to ensure that protected areas are representative of all biomes and ecosystems (UNEP, 2010). In Brazil, for example, most protected areas (77%) are within the Amazon region, which reflects the successful implementation of the ARPA programme and the government's efforts to reduce deforestation in the region. While the Amazon will remain important, greater effort is needed in the Cerrado and Caatinga biomes where protection is low and higher deforestation is anticipated. In Chile, only 11 of 64 sites identified as national protection priorities are fully or partially within the boundaries of official protected areas, 46% of the country's wetlands are protected, and coastal protection near growing population centres is limited. In Peru, only 12 of the country's 21 terrestrial eco-regions are represented (OECD/ECLAC, 2017, 2016; OECD, 2015). Priority should therefore be given to under-represented ecosystems as countries continue to expand their protected area systems.

Declaration of protected areas alone is not enough if it is not linked to effective measures or management. However, effective management is a significant challenge. Despite legal requirements, many protected areas operate without a management plan; those that do cannot always ensure that it is implemented and that biodiversity is effectively conserved or used sustainably within the protected zone. This is largely due to human and financial capacity constraints. In Chile, for example, most protected areas lack sufficient resources, including park rangers and a monitoring system (OECD/ECLAC, 2016). In Brazil, less than half of protected areas in the Amazon biome had an approved management plan in 2012, even though a management plan is a precondition for sustainable use (e.g. for tourism, sustainable logging or use by the local community). In Colombia, 93% of terrestrial protected areas have adopted management plans, but endemic and threatened species are not always adequately protected (OECD/ECLAC, 2014). Peru has increased the attention given to Protected Natural Areas (PNA): while in 2003, 33 out of 40 PNAs were staffed and only 17 had management plans, in 2015, 61 out of 64 PNAs were staffed and 41 had management plans (OECD/ECLAC, 2017). Brazil has also made progress in the management of its protected areas, but limited resources and capacity have constrained implementation (Box 4.2). Strengthen the effective management of protected areas in Latin America will require adequate funding as well as strengthened governance schemes with the participation of local communities.

Box 4.2. Brazil improves governance and management of protected areas

The law establishing Brazil's National System of Protected Areas (SNUC) introduced several features that helped improve the governance and management of protected areas. One such feature is the requirement to establish management committees that would facilitate the involvement of local communities and stakeholders in decisions concerning protected areas. The committees include government officials and representatives of the private sector and civil society.

The law requires protected areas to establish management plans within five years of their creation. Plans are a condition for sustainable public use (such as tourism, environmental education and sustainable logging) and local community resource use (e.g. harvesting, fishing, farming). However, many areas have not met this deadline due to limited resources and capacity. In 2012, only 94 of 247 protected areas in the Amazon biome had an approved management plan.

The law also introduced instruments for managing protected areas at a landscape scale, allowing for connections among and within ecosystems and recognising the importance of ecological corridors to maintain ecological processes. It introduced the opportunity to integrate multiple protected areas into a "mosaic" if they are in proximity or overlap. This approach allows for the development of shared solutions to issues such as land and resource use in border zones, access to protected areas, enforcement, monitoring and evaluation of management plans, and research. As of 2014, 14 mosaics had been approved.

Source: OECD (2015), *OECD Environmental Performance Reviews: Brazil 2015*.

4.2.2. Biodiversity conservation on private land

Private protection initiatives can be particularly important for improving representativeness in priority ecoregions, as often the greatest pressures are near populated centres and in agricultural regions where land is privately owned.

Biodiversity conservation on private land can be achieved through both regulatory requirements and voluntary engagement. Brazil's Forest Code, for example, requires landholders to set aside a share of their private land for conservation. In the Amazon, land holders have to preserve 80% of forested land on their private property, while land owners in other regions generally have to preserve 20% of native flora. An innovative system of tradable forest quotas (see Section 4.3) aims to facilitate compliance with this requirement.

In other countries, private voluntary donations play an important role for biodiversity conservation on private land. There are a few examples where individuals, non-profit organisations or companies independently decided to purchase, donate or set-aside a portion of land for conservation. This was the case with the Pumalin Park in Chile, one of the largest private protected areas in the world (Box 4.3). Tax incentives, subsidy programmes or other support measures can stimulate such donations and help develop a culture of environmental philanthropy. In Chile, pending legislation will allow for private initiatives to be brought into the official protected area system, financing of management plans and incentives for further private conservation efforts (OECD/ECLAC, 2016).

Box 4.3. Chile's Pumalin Park, one of the largest private protected areas

Pumalin Park was originally created by the founder of the American clothing company *The North Face*, Douglas Tompkins in 1991. As a regular visitor to southern Chile fond of skiing, kayaking and hiking, he decided to purchase 17 000 ha to protect primeval native temperate rainforest at risk of logging. The park has grown over time, acquiring an additional 230 000 ha and establishing a network of campgrounds, trails, information centres and other public facilities. Chile has designated the park a nature sanctuary and the lands have been donated to a Chilean Foundation – Fundación Pumalín – for the administration and ongoing preservation of the park.

Source: OECD/ECLAC (2016), *OECD Environmental Performance Reviews: Chile 2016*.

4.2.3. Other regulatory approaches

Many Latin American countries also use other regulatory approaches such as standards, licensing, permitting, and planning tools to promote biodiversity conservation and sustainable use. Regulatory restrictions on activities potentially harmful to biodiversity are common across Latin American countries. Mexico, for example, places restrictions on: whale watching activities; sea turtle, shark and stingray fishing; and the use of gill nets (OECD, 2013a).

Most countries in Latin America require Environmental Impact Assessments (EIAs) of major projects, though controversy remains in many countries regarding their effectiveness in protecting biodiversity. EIAs are structured processes for obtaining and evaluating the potential environmental impacts of a project prior to decision-making. They are usually applied to proposed major projects, such as power stations or mines (UNEP, 2004). In Chile, the EIA process has historically dealt with biodiversity issues in an ad hoc manner, leading to an uneven treatment of projects. It also tends to come too late in the project design process to result in significant change, does not provide avenues for adjustment once the project is operating, and does not cover smaller projects such as small mines that can have important impacts on ecosystems (OECD/ECLAC, 2016). Many of these issues are not unique to Chile, however. Several OECD countries do not require EIAs for smaller projects and one of the key challenges in most countries, such as France, is involving the public early enough in the project to be able to make meaningful changes (OECD, 2016a). The development and use of technical guides can facilitate the full consideration of the impact on land and marine biodiversity in EIA processes.

Strategic Environmental Assessment (SEA) is increasingly being used in Latin American sectoral policies and land-use plans, though not yet comprehensively or consistently. SEAs incorporate a range of analytical and participatory approaches to integrate environmental considerations into policies, plans and programmes and evaluate the inter-linkages with economic and social considerations. Colombia has promoted the use of SEAs in sectoral policy development, but has not made it a legal obligation (OECD/ECLAC, 2014). SEA has also been increasingly in Chile and Peru. In Chile, most territorial plans are required to undergo an SEA, although less than half of them do. Mexico and Brazil have no legal requirements for SEA.

Land-use planning and zoning can be another effective tool for biodiversity conservation and sustainable use. Brazil, for example, put in place a National Environment Policy for environmental and ecological-economic zoning (ZEE) aimed at allocating compatible activities in defined environmental areas to maintain sustainable use of natural resources and a balanced ecosystem. Several states have also developed ZEEs, and the 2012 Forest Code requires all states to approve their ZEEs by 2017. While the maps and guidelines are useful tools for territorial and development planning, further work is needed to improve their effective use in spatial planning and policy making, and bolster the capacity at the municipal level to implement zoning requirements (OECD, 2015). Peru is also using ZEEs as one of its land-use planning tools, and 13 of 24 regions have approved ZEEs (OECD/ECLAC, 2017). Colombia advanced in the protection of 2 million ha of moorlands and 1.8 million ha of wetlands in trough zoning processes that prioritised environmentally critical areas that were not under any type of legal protection.

4.3. Economic instruments

Economic instruments are important tools for promoting biodiversity conservation and sustainable use efficiently, while also offering the potential to raise revenue. Latin American countries are increasingly adopting economic instruments, but many are in the early stages of development or are not yet sufficiently stringent to significantly impact biodiversity outcomes.

4.3.1. Payment for ecosystem services programmes

Latin American is a leader in the use of payments for ecosystem services (PES). PES are based on the recognition that well-functioning ecosystems provide important services essential for the economy and human well-being, such as reliable and clean flows of water and productive soil. PES are agreements whereby a user or beneficiary of an ecosystem service pays individuals or communities whose management decisions influence the provision of ecosystem services (OECD, 2010). Some countries have combined PES systems with social objectives, helping to provide financing to impoverished rural communities for their involvement in biodiversity conservation and sustainable use.

Costa Rica was an early pioneer in the use of programmes. The country developed the first PES programmes over 20 years ago, paying land owners to protect forests in return for their ecosystem benefits (such as conserving species, regulating river flows and storing carbon). Forest cover has returned to over 50% of the country's land area (from a low of 20% in the 1980s) largely as a result of the PES system and a ban on land-use change in forests (Barton, 2013). Approximately 1 million ha of forest in Costa Rica has been part of the PES programme. The current programme favours indigenous territories, areas with low social development scores and properties under 50 ha. The system is also helping to formalise land tenure and to update property registers needed to collect taxes.

Mexico now runs one of the world's largest national PES programmes, covering more than 3.25 million ha. It comprises two PES initiatives involving forest management, which were brought together under the same umbrella in 2011: the Hydrological Ecosystem Services Programme (PSAH) launched in 2003 and the Programme to Develop Ecosystem Service Markets for Carbon Sequestration and Biodiversity and Improved Agro-forestry Systems (CABSA) launched in 2004. The PSAH is funded by a national fee on water use, while the CABSA's budget is renegotiated by congress every year and therefore does not have stable, long-term funding. Ecosystem service providers are predominantly *ejidos* – areas of communal land used for agriculture. Verification of forest cover is done through satellite image analysis or ground visits. The programme also includes an environmental endowment fund and the promotion of local PES mechanisms through matching funds. An important feature of the PES programme is that it targets areas with high biodiversity benefits, high risk of loss and low opportunity cost (OECD, 2010). Mexico adjusted and revised its programme several times to take into account the first two of these elements. The conservation impact of the PSAH initiative has been fairly low, and could be improved by putting greater weight on environmental eligibility criteria (OECD, 2013a).

While Brazil does not yet have a national legal framework governing PES, several states and municipalities have developed their own laws and PES programmes. Brazil has implemented several PES and conditional cash-transfer programmes at the federal and state levels, including Bolsa Verde and Bolsa Floresta – which provide payments to extremely poor households in rural forest communities to compensate them for conservation activities (Box 4.4). The beneficiaries of the programmes are mainly rural family producers and settlers, traditional communities and indigenous peoples, with financing generally provided by governments. A national PES framework in Brazil could help standardise programmes, improve monitoring and effectiveness and lower transaction costs (OECD, 2015).

Box 4.4. Linking PES with social protection: Bolsa Verde and Bolsa Floresta

Bolsa Floresta, launched in 2007 by Amazonas state, provides monthly cash payments of about USD 20 to families living in protected areas in exchange for forest conservation efforts (e.g. for limiting the amount of forested land converted for farming). Bolsa Floresta was the first of its kind and became of the world's largest PES programmes, reaching more than 35 000 people in 15 protected areas in 2013.

Building on this initiative, the federal government launched Bolsa Verde in 2011 as part of the broad anti-poverty programme Brasil sem Miséria. The programme provides payments for adoption of environmental practices and technical training to support beneficiaries in meeting their conservation commitments. It is seen as a potentially efficient way to curb deforestation, with low payments per hectare of avoided deforestation. However, implementation is complex and complementary training activities are insufficiently developed. Developing monitoring mechanisms and ensuring a link with the national land registration programme (the Rural Environmental Cadastre) would help improve effectiveness and reduce management costs for Bolsa Verde and the existing PES programmes.

Source: OECD (2015), *OECD Environmental Performance Reviews: Brazil 2015*.

In Colombia, legislation was extended in 2007 to support PES programmes and a law on PES was enacted in 2017. PES programmes at the national level include the Forestry Incentive Certification Programme for commercial reforestation and the Forest Ranger Families Programme which helps shift families from growing illicit crops towards conservation, restoration and legal and sustainable production. There are also several sub-national programmes, mainly focused on watershed conservation and restoration, including for Colombia's capital city Bogotá (Box 4.5) (OECD/ECLAC, 2014).

Box 4.5. Chingaza National Park in Colombia values ecosystem services from the Páramo

The páramo, or high Andean moorland, provides crucial ecosystem services, such as regulating the quantity and quality of water. Around 70% of the Colombian population's water supply originates from upland areas. Chingaza National Park in the páramo is the main source of water supply for the 8 million inhabitants of Bogotá, while also supplying water for hydropower generation. Conservation measures in the Park decrease the generation of sediment in the water, securing the quality of water and reducing the costs of water treatment.

The Bogotá water utility makes a voluntary annual payment to Chingaza National Park, and is also charged a water use fee that includes a minimum charge plus a variable component that reflects the investment needs for conservation, the socio-economic circumstances of stakeholders and the scarcity of water resources. Detailed fee calculation is possible through the capacity of the national park to gather technical data.

The regime is a model of a successful "payment for ecosystem services" approach. However, revenue raised does not reflect the full value of the ecosystem service and there continues to be political resistance to fully deploying the fee system as fears of increased costs could damage economic activity in the region and affect poor households.

Source: OECD/ECLAC (2014) *OECD Environmental Performance Reviews: Colombia 2014.*

As Latin American countries develop, expand and update PES systems, they could look to lessons learned as well as additional features such as those used in Mexico to prioritise areas with high biodiversity benefits and/or high threat of loss. Most EPRs advise countries to carefully monitor and evaluate the effectiveness of the PES schemes. Adverse auctions mechanisms could be considered used to improve the cost-effectiveness of PES systems. The focus of PES systems on providing financing to impoverished rural communities may help smooth approval and implementation of PES programmes, yet countries should be careful to ensure it does not erode environmental benefits.

4.3.2. Biodiversity offsets

Biodiversity offsets are conservation actions designed to compensate for significant, residual biodiversity loss from development projects after reasonable steps have been taken to avoid and minimise biodiversity loss at a development site (BBOP, 2009). Biodiversity offsets are economic instruments based on the polluter pays approach. The most common objective adopted in offset programmes is to deliver No Net Loss (e.g. of a habitat, species, ecological status, ecosystem services), although several programmes around the world have adopted a more ambitious goal of Net Gain (OECD, 2016b).

At least 56 countries around the world have laws or policies that specifically require biodiversity offsets or some form of compensatory conservation for particular impacts (OECD, 2016b). In Latin America, several countries have biodiversity offset systems in place, including Mexico, Brazil and Colombia, and Chile is working to develop a national offset programme.

Mexico adopted its Forest Land Use Change Compensation mechanisms in 2005. Under the mechanisms, successful land-use change applicants are required to reforest an area at least the same size as the deforested area with species of the same type. The developer can choose whether to create its own offset or pay into a compensation fund at a compensation ratio greater than 1:1. However, the current system has not assessed whether the reforestation activities linked to compensation are successful and whether their location and timing truly compensate for the environmental services lost (OECD, 2013a).

Brazil has a number of offset mechanisms in place. One is integrated into its environmental licensing procedure, whereby project developers can be required to pay compensation based on the severity of the environmental impact of the project. However, clear mechanisms to monetise the environmental impact and the amount of compensation are needed. The revenue generated is earmarked for protected areas. Another is the land offset mechanism and tradable forest quotas introduced by the Forest Code. Landholders that are not compliant with forest set-aside requirements can make up for this deficit either by buying private property within official protected areas on behalf of the government, which allows the consolidation of protected areas, or by purchasing Environmental Reserve Quotas (see section below on Fishing and Forestry Tradable Quotas and Fees). However, for this mechanism to operate, Brazil's Rural Environmental Cadastre will need to be fully implemented (OECD, 2015).

Colombia developed a manual in 2012 that provides guidance on how the impact on ecosystems from development projects can be offset by the developer providing an equivalent form of ecological compensation. The 2014 OECD EPR of Colombia highlighted the positive step in developing the manual, while noting that effective implementation would require enforcement, consistent application of the requirements across sectors and regions, and effort to ensure that the offsets are additional to what would have taken place anyway (OECD/ECLAC, 2014).

Chile's use of biodiversity offsets is at a very early stage. The Ministry of Environment and the Environmental Assessment Service have released a guide on biodiversity offsets as compensatory measures in EIA. The proposed legislation creating the Biodiversity and Protected Area Service will provide the legal framework for establishing biodiversity offsets and biobanks of certified and quantified conservation initiatives (OECD/ECLAC, 2016).

While biodiversity offsets have significant potential to improve overall outcomes, policies in place to date have had mixed results in terms of environmental effectiveness. Careful design and implementation of the instrument is key to success. Experience from some of the regimes that have been in place the longest – such as wetland banking in the United States – will be helpful to Latin American countries as they develop and refine their own systems. The OECD publication *Biodiversity Offsets: Effective Design and Implementation* offers good practice insights as well as case studies from the United States, Germany and Mexico. Germany has over 1 000 biobanks – where developers can purchase credits from a repository of existing offsets – operating or under development. The US species mitigation offset system has at least 143 different credit types, with 92 for species and 51 for habitat (OECD, 2013b).

4.3.3. Water markets and charges

While there are some water markets and charges in place in Latin America, most do not yet explicitly reflect water needs for biodiversity across ecosystems. Effective water markets or water use charges will be increasingly important in Latin America as water scarcity concerns grow. Water prices that reflect the scarcity or vulnerability of the resource can promote reduced and more efficient water use. It is important, however, that the water needs of biodiversity and ecosystems are considered in the design of instruments, along with the needs of agriculture, industry and municipalities.

In Colombia, environmental legislation supports the financing of watershed management. Hydroelectric plants must transfer 6% and thermal energy plants 4% of their revenue to regional and municipal authorities to carry out watershed conservation and sanitation projects, with over USD 80 million raised annually. Entities constructing or operating irrigation projects or other water abstractions are required to use 1% of the amount invested to pay for watershed protection. All water users are also required to pay a fee, which raises over USD 10 million annually. Departmental and municipal governments are mandated to spend 1% of current income to purchase or manage lands that protect municipal water sources or for payment for ecosystem services, with the legislation allowing for collaboration across districts (OECD/ECLAC, 2014). Colombia recently raised water fees by over 100% for the industrial, mining and hydrocarbons sectors.

The 2014 EPR of Colombia noted that while the system had significant potential, the fees were too low to finance both water service provision and watershed protection, and more effort is required to improve the rate of water fee collection, estimated at 67.5% in 2010. The consolidation of water fee revenues into three water funds is seen as an important measure to enhance the efficiency of spending, allowing for supplementary funding from donors and international financial institutions. The funds are also managed by a committee of stakeholders. The water fund model has been adopted in several Latin American countries, and the Latin American Water Funds Partnership is working to scale up the use of water funds in the region (OECD/ECLAC, 2014).

Chile has long had a market of water use rights, but existing user rights do not allow for meeting minimum flow requirements in half of the river basins in arid northern Chile. This is due to insufficient regulation and transparency of the water market, which have led to over-allocation and extreme concentration of water rights and overexploitation of some aquifers. Rising water tariffs encouraged urban households to reduce consumption of drinking water by 18% between 2000 and 2013, but the amount of water abstracted for public water production increased by 23% as a result of rising water losses from leaking infrastructure (OECD/ECLAC, 2016).

Brazil's National Water Resources Policy Law introduced water abstraction and effluent charges as water resource management tools, but only a few states and river basins charge for water and those that do have fees too low to influence use. Hydropower plants are required to provide financial compensation of 6.75% of the value of electricity produced, and mines between 1-3% of turnover, to compensate for the use of water and natural resources, but revenue is not earmarked to environmental activities as in Colombia (OECD, 2015).

4.3.4. Fishing and forestry tradable quotas and fees

Tradable quota systems and fees can also provide efficient mechanisms to promote sustainable use of biodiversity. Successful implementation requires careful design that reflects biodiversity priorities and adequately finances enforcement and monitoring.

Chile introduced a quota system for its fishing industry in 2001, with quotas distributed between industrial and small-scale sectors, and a tradable quota licence system for industrial fisheries. These reduced fish catches by 64% between 2004 and 2013. However fish production from aquaculture almost tripled over 2000-12, and the effluent, pesticides and medicines this generates are a major source of pollution of, and pressure on, inland waters, estuaries and marine ecosystems (OECD/ECLAC, 2016).

Colombia has had a forestry fee in place since 1982, initially set as 10% of the value of wood extracted. As of 1993, regional authorities are able to set their own fees. The system is challenged by the fact that illegal logging, which does not contribute to the fees, still accounts for a large percentage of timber produced in Colombia (OECD/ECLAC, 2014).

Brazil's Forest Code – updated in 2012 – introduced an economic instrument to facilitate compliance with land set-aside obligations that require landowners to maintain native vegetation on a proportion of their properties and along water bodies and sensitive areas. Environmental Reserve Quotas (CRAs) can be issued for each hectare of area maintained as native vegetation in excess of the requirement. This quota can then be purchased to offset a deficit in a different property in the same biome, and preferably in the same state. The 2015 OECD EPR highlighted the promise of the initiative, but noted that care would need to be taken to ensure conservation of areas with high biodiversity value rather than only areas with low opportunity cost (OECD, 2015).

4.3.5. Financial incentives and subsidies to promote sustainable use of biodiversity and ecosystems

Financial incentives can be an effective approach to encourage conservation, sustainable use and restoration activities. In areas where high biodiversity benefits overlap with areas where the poor are concentrated, such incentives can also provide social co-benefits that make them more appealing to adopt.

Many of the biodiversity-related economic instruments applied in Mexico are subsidy-based. Over 53% of the forests are owned by local or indigenous communities that are generally poor. A national reforestation programme (PRONARE) supports landowners or users reforesting degraded forest land by providing seedlings, training and funding. This has succeeded in reforesting a much larger land area than what would have been without the programme (UACH, 2010). Another programme promotes the conservation and sustainable use of wildlife, income generation and employment through Management Units for Wildlife Conservation (UMAs) and Facilities for Wildlife Handling (PIMVS) in rural areas. Sustainable fishing is promoted with buyouts for fishers who are willing to accept payments to stop fishing or switch to alternative methods that help protect the vaquita, one of the world's smallest and most at-risk cetaceans. These measures are believed to have helped reduce threats to conservation of vaquitas and to have begun decreasing the total level of fishing, with conservation benefits for other marine species (OECD, 2013a).

Chile also provides subsidies for native forest conservation, and is establishing a National Biodiversity Fund to implement new economic instruments and finance conservation programmes outside protected areas (OECD/ECLAC, 2016). Peru provides direct transfers to indigenous and rural communities for forest conservation (OECD/ECLAC, 2017).

Brazil developed an innovative incentive to improve enforcement and compliance with deforestation requirements. Municipalities with critical deforestation levels are placed on a blacklist maintained by the Ministry of Environment. Public financial institutions then restrict credit available to those on the blacklist. The Brazilian Central Bank has also made access to subsidised rural credit in the Amazon biome conditional on the legitimacy of land claims and provision of information to demonstrate compliance with environmental regulations (OECD, 2015).

Green public procurement can also support the development of sustainable domestic industries. As part of its 2012 Sustainable Procurement Policy, Brazil's government launched a national procurement programme targeting biodiversity-related products. Similar initiatives also exist at the state level. The 2009 National Plan to Promote the Production Chain of Socio-Biodiversity Products (PNPSB) provides facilitated access to credit and markets as well as technical assistance. A minimum price is provided for select products, such as açai fruit, natural rubber and Brazil nuts. While this price support only benefits a small fraction of total production of the targeted products, production and commercialisation of socio-biodiversity products and competition among buyers has increased (OECD, 2015).

4.3.6. Removing environmentally-harmful subsidies

Subsidies that are counter-productive to efforts to address pressures facing biodiversity remain in many countries across Latin America. Examples include government subsidies for agricultural production and irrigation that do not include environmental criteria, subsidies for small mine production, subsidies for non-native forestry plantations, and tax exemptions for fertilisers and pesticides. Such subsidies stimulate increased production and input use, which puts pressure on the natural resource base and on biodiversity. For example, chemical fertilisers and pesticides have negative impacts on soil and water quality, and can harm human health and ecosystems. Removing or reforming these subsidies can help re-align incentives towards more biodiversity-friendly approaches, such as the more efficient use of inputs.

Brazil, for example, provides tax reductions to certain fertilisers and pesticides, as well as market price support through guaranteed minimum prices and direct government purchases which encourage increased production (OECD, 2015). In Chile, Mexico and Colombia producer support is also linked to input use. In Chile, fertiliser and pesticide use has increased faster than total agricultural production, indicating inefficient use (OECD/ECLAC, 2016). In Mexico, the value of subsidies supporting conventional production is far higher than that of environment-oriented programmes such as the PES system (OECD, 2013a). Low or subsidised water prices are a barrier to efficient water use in all five EPR countries (Section 5.2).

Fiscal reform need not be detrimental to agriculture sectors. For example, Denmark mobilised DKK 461 million (roughly USD 69 million) in 2010 from its pesticide tax, with 60% of the tax revenue channelled back to the agriculture sector through different subsidy schemes. Indonesia removed its pesticide subsidy, and three plantings later had record levels of rice production combined with savings of over USD 100 million (OECD, 2013b).

In Chile, the Biodiversity Finance Initiative (Biofin) co-ordinated by the United Nations Development Programme in co-operation with finance and environment ministries identified two subsidies harmful to biodiversity: support to irrigation infrastructure, and subsidies to small-scale mining. Current irrigation subsidies do not incorporate environmental criteria, and have allowed the drainage of wetlands and replacement of areas of rich biodiversity with monocultures. Mining uses large amounts of water, and mining waste (tailings) contaminate soil, surface water and groundwater, hence damaging ecosystems. Subsidies for forest plantations have also encouraged replacing native forests with exotic species, which absorb significant groundwater and reduce biodiversity (OECD/ECLAC, 2016).

4.4. Voluntary and information instruments

Voluntary and information instruments can be important complementary tools to regulatory and economic instruments. They are particularly useful in areas where economic and environmental objectives align. For example, environmental certification and labelling can help companies compete in global markets increasingly demanding environmentally-friendly products (OECD, 2013b). Private sector agreements can offer reputational benefits for companies while attaining biodiversity objectives.

4.4.1. Certifications and eco-labelling

Chilean companies have realised the economic benefits of certifications and eco-labels, given growing demand from international consumers for sustainable production methods from suppliers, particularly in forestry, aquaculture and agriculture. However, the variety of international and independent eco-labels in the Chilean market tends to confuse consumers (OECD/ECLAC, 2016). Mexico has two sustainable forestry certifications, two green coffee certifications, and eco-certification for tourism-related businesses (OECD, 2013a). Colombia has implemented a national green labelling system – Sello Ambiental Colombiano – that aims to increase the proportion of goods and services with environmental certification to 30% by 2019. Environmental certifications are also in place for tourism, wine, coffee, organic produce and other goods and services. Brazil also has a national green labelling system, but this has not yet managed to co-ordinate and better articulate environmental labelling initiatives (OECD, 2015).

4.4.2. Private sector agreements

In Brazil, the business sector has been motivated to help combat desertification in the Amazon by what is referred to as the "Soya Moratorium". In 2006, following a report from Greenpeace and pressure from consumers, a large group of companies (including McDonald's and Wal-mart), in co-operation with the Ministry of Environment, agreed to stop using soy grown on cleared forestland in the Brazilian Amazon. The initiative was one of the first voluntary zero-deforestation agreements in the world. In 2004, nearly 30% of soya expansion occurred through deforestation. By 2014, the rate was reduced to 1% in the Amazon. A similar initiative – the Beef Slaughterhouse Pact – was developed for Brazil's cattle industry (OECD, 2015). Further expansion of such initiatives to other regions and sectors could provide additional biodiversity benefits.

Chile has used its Clean Production Agreements with industry to set specific targets and promote action in exchange for financial support. While the focus to date has been on energy, waste and water use, there is work underway to incorporate biodiversity objectives. An agreement with the fruit sector, for example, aims to reduce the impact of pesticides on pollinators (OECD/ECLAC, 2016). The Colombian government has signed Zero Deforestation Agreements with private companies from the oil palm, beef, milk and wood sectors, in which companies commit to not generate negative impacts on forests and other strategic ecosystems.

References

Barton, D. N. (2013), *Payment for Ecosystem Services: Costa Rica's Recipe*, International Institute for Environment and Development, London, www.iied.org/payments-for-ecosystem-services-costa-rica-s-recipe.

BBOP (2009), *Business, Biodiversity Offsets and BBOP: An Overview*, Business and Biodiversity Offsets Programme, Washington, D.C., www.forest-trends.org/documents/files/doc_3125.pdf.

Juffe-Bignoli, D., et al. (2014), *Protected Planet Report 2014*, UNEP-WCMC: Cambridge, UK, www.unep-wcmc.org/system/dataset_file_fields/files/000/000/289/original/Protected_Planet_Report_2014_01122014_EN_web.pdf?1420549522.

OECD (2018), "Biodiversity: Protected areas", *OECD Environment Statistics* (database), https://doi.org/10.1787/5fa661ce-en (accessed on 16 July 2018).

OECD (2016a), *OECD Environmental Performance Reviews: France 2016*, OECD Publishing, Paris, http://dx.doi.org/10.1787/9789264252714-en.

OECD (2016b), *Biodiversity Offsets: Effective Design and Implementation*, OECD Publishing, Paris. http://dx.doi.org/10.1787/9789264222519-en.

OECD (2015), *OECD Environmental Performance Reviews: Brazil 2015*, OECD Publishing, Paris, http://dx.doi.org/10.1787/9789264240094-en.

OECD (2013a), *OECD Environmental Performance Reviews: Mexico 2013*, OECD Publishing, Paris, http://dx.doi.org/10.1787/9789264180109-en.

OECD (2013b), *Scaling-up Finance Mechanisms for Biodiversity*, OECD Publishing, Paris, http://dx.doi.org/10.1787/9789264193833-en.

OECD (2010), *Paying for Biodiversity: Enhancing the Cost-Effectiveness of Payments for Ecosystem Services*, OECD Publishing, Paris, http://dx.doi.org/10.1787/9789264090279-en.

OECD/ECLAC (2017), *OECD Environmental Performance Reviews: Peru 2017*, OECD Publishing, Paris, http://dx.doi.org/10.1787/9789264283138-en.

OECD/ECLAC (2016), *OECD Environmental Performance Reviews: Chile 2016*, OECD Publishing, Paris, http://dx.doi.org/10.1787/9789264252615-en.

OECD/ECLAC (2014), *OECD Environmental Performance Reviews: Colombia 2014*, OECD Publishing, Paris, http://dx.doi.org/10.1787/9789264208292-en.

UACH (2010), "Informe de evaluación externa de los apoyos de reforestación, ejercicio fiscal 2009", *Universidad Autónoma de Chapingo*, Comisión Nacional Forestal, Texcoco, México, http://148.223.105.188:2222/gif/snif_portal/administrator/sistemas/evaluaciones/1301593358_2009_reforestacion_resumen_eje.

UNEP (2010), *State of Biodiversity in Latin America and the Caribbean*, United Nations Environment Programme, Panama City and Nairobi, www.unep.org/delc/Portals/119/LatinAmerica_StateofBiodiv.pdf.

UNEP (2004), *Environmental Impact Assessment and Strategic Environmental Assessment: Towards an Integrated Approach*, United Nations Environment Programme, Geneva, www.unep.ch/etu/publications/textONUBr.pdf.

5. Financing

This chapter examines how the reviewed Latin American countries finance the implementation of their biodiversity conservation strategies and action plans. It provides an overview of public budget allocations, private sector contributions and the role of international development assistance. It presents examples of conservation trust funds, which are commonly used in the reviewed countries.

The statistical data for Israel are supplied by and under the responsibility of the relevant Israeli authorities. The use of such data by the OECD is without prejudice to the status of the Golan Heights, East Jerusalem and Israeli settlements in the West Bank under the terms of international law.

5.1. Introduction

The first CBD High-Level Panel report in 2012 estimated that between USD 150 billion and USD 440 billion per year would be required globally to meet the Aichi Biodiversity Targets by 2020. They also noted, however, that the benefits secured through implementing the Aichi Targets are likely to significantly outweigh costs (CBD High-Level Panel, 2014).

The overall trend in biodiversity financing in Latin America is upward, with most government budgets growing between 2000 and 2015. Countries such as Chile and Mexico have managed to increase rates of private revenue generation for protected area funding, and countries such as Brazil are expanding their use of biodiversity funds. International finance also remains important, with Latin America receiving more development assistance for biodiversity-related activities than other region. However, the five EPRs undertaken for Brazil, Chile, Colombia, Mexico and Peru show that overall financing remains inadequate to achieve country and regional biodiversity objectives. Protected area management, enforcement of existing environmental laws and ecosystem monitoring and reporting are particularly impacted by the lack of financial resources. While domestic public financing is essential to conservation efforts, revenue raised from protected area entrance fees and other economic instruments as well as international financing can also play an important role.

A 2010 study by the UNDP and The Nature Conservancy of the financial sustainability of Protected Areas in Latin America found a financing gap of USD 314 million per year just to meet basic management needs. More rigorous management would require USD 700 million per year. The study also cautioned that the situation could get worse, with growing funding needs to respond to pressures and increased commitments, and risks to financing that is not stable or secure. There is, however, considerable variability across countries in spending levels. For example, it was estimated that Chile, Peru and Brazil only have roughly half of their financial needs for protected areas covered (Bovarnick et al., 2010).

The Biodiversity Finance Initiative (Biofin), launched in 2012 by the UNDP, aims to support countries to define biodiversity finance needs and gaps through detailed national assessments. Brazil, Chile, Colombia, Mexico and Peru are all participating in the programme. The initiative provides a methodology to enable countries to measure their current biodiversity expenditures, assess their financial needs in the medium term and identify the most suitable finance solutions to bridge their national biodiversity finance gaps (Biofin, 2016).

5.2. Domestic public financing

Latin American countries continue to rely mainly on domestic public financing to implement biodiversity conservation strategies and action plans. While budgets have generally increased over the last 10-15 years, overall financing is not sufficient to attain stated biodiversity objectives. Most Latin American countries are spending under USD 5 per hectare of protected area, which is below spending by OECD peers such as Australia, Sweden and the United States (Figure 5.1). Costa Rica and Argentina are the exceptions. In addition to the total volume of funding, analysis about the breakdown of spending could help identify where funding gaps are largest, and better channel available resources to priority areas for action.

Figure 5.1. Protected area funding is uneven

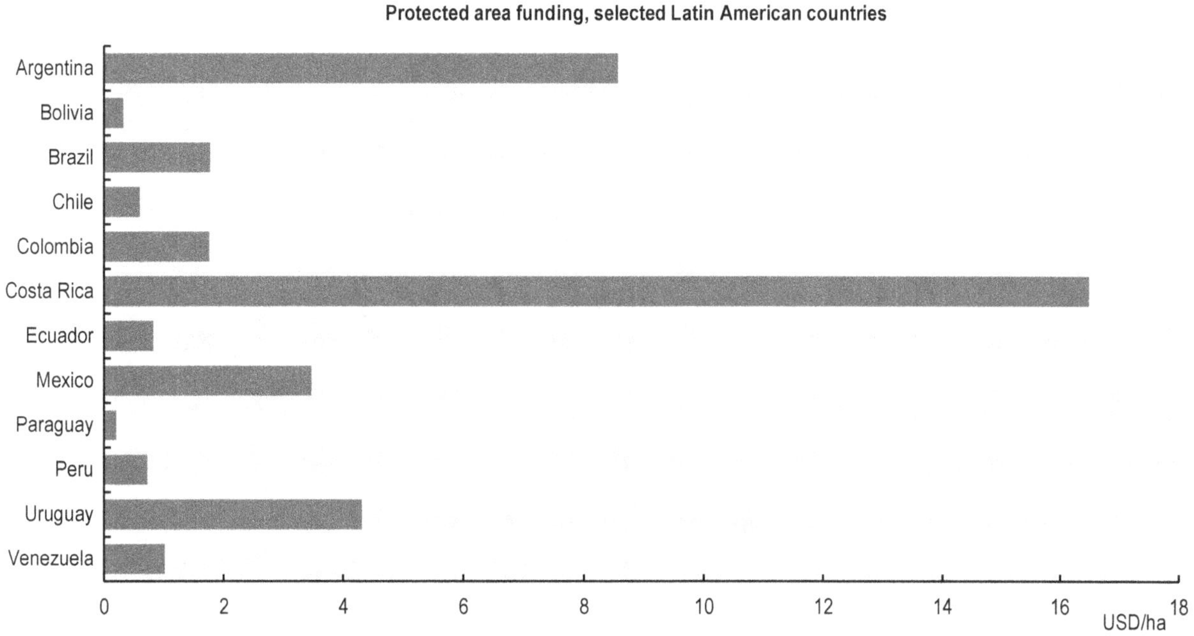

Protected area funding, selected Latin American countries

Country	USD/ha
Argentina	~8.5
Bolivia	~0.3
Brazil	~1.8
Chile	~0.6
Colombia	~1.8
Costa Rica	~16.5
Ecuador	~0.8
Mexico	~3.5
Paraguay	~0.2
Peru	~0.7
Uruguay	~4.3
Venezuela	~1.0

Source: Bovarnick et al. (2010), *Financial Sustainability of Protected areas in Latin America and the Caribbean: Investment Policy Guidance*; Mansourian and Dudley (2008), *Public Funds to Protected Areas*.

StatLink https://doi.org/10.1787/888933886113

Chile has significantly increased financing for biodiversity from public resources, with the budget allocation growing by 176% between 2000 and 2014, slightly faster than total central government outlays for environmental protection (+174%) and more than the total government budget (+139%). Biodiversity accounted for the largest share of all estimated environmental protection expenditure in 2012 (28%) and 0.26% of the 2014 central government budget. However, Chile's funding per hectare of protected area is among the lowest in Latin America (Figure 5.1). Building on the current USD 41 million per year, it is estimated that an additional USD 35 million per year would be needed to finance an improved protected area system with effective management and monitoring. Proposed legislation establishing a new Biodiversity and Protected Area Service includes a request for an increase in public financial resources to approximately USD 47 million per year. The proposed consolidation of previously fragmented agencies involved in biodiversity activities also has the potential to improve the efficiency and effectiveness of public spending (OECD/ECLAC, 2016).

In Peru, public funding for biodiversity rose by 500% between 2004 and 2010, but a study by the Universidad del Pacifico identified an annual shortfall of roughly USD 35 million. Increases in financing for environmental enforcement in Peru are, however, a positive development (Box 5.1).

> **Box 5.1. Peru increases financing for environmental enforcement**
>
> Peru has strengthened environmental enforcement, through OEFA, the lead body in the National Environmental Assessment and Oversight System. OEFA supervises compliance with environmental regulations in four sector groupings: medium and large-scale mining; hydrocarbons and electricity; commercial fisheries and aquaculture; and the brewery, papermaking, cement and tannery industries.
>
> The government increased the environmental enforcement budget from USD 16 million in 2012 to USD 71 million in 2015, and increased the maximum level of fines for noncompliance threefold. The additional financing allows for a significant increase in the direct auditing of firms, as well as supervision of other environmental enforcement entities.
>
> *Source*: OECD/ECLAC (2017), *Environmental Performance Review of Peru 2017*.

While Brazil does not have comprehensive or consistent information on public and private biodiversity-related spending, federal budget outlays for biodiversity-related programmes grew by 50% in real terms between 2010 and 2014. The Chico Mendes Institute for Biodiversity Conservation (ICMBio) is the main institution for biodiversity-related programmes, and its budget increased by 57% between 2008 and 2014 (OECD, 2015).

Mexico increased expenditure on biodiversity from MXN 2.6 billion in 2001 to MXN 8.4 billion in 2009 (approximately USD 135 million to USD 446 million). The public sector is the greatest contributor to conservation projects, with one study estimating approximately 74% of funding for conservation projects came from public sources (OECD, 2013).

5.3. Private revenues

With public financing falling short of what is needed, it will be increasingly important for Latin American countries to maximise other sources of financing, including from the private sector and NGOs. Protected area entrance fees, licenses and fees for tourism and other activities in protected areas, permits for research, concessions, payment for ecosystem services programmes (Section 4.3), and instruments for private sector support are all used in Latin America, but there remains scope to increase the scale and coverage of these instruments.

Growth in nature-based tourism offers an opportunity to boost biodiversity financing in Latin America. In 2010, private revenues from protected areas (e.g. from access fees and concessions) represented only 10% of total protected area funds available in the region. Chile and Mexico have, however, managed to achieve rates above 25% (Bovarnick et al., 2010). In Brazil, only 17% of protected areas that could receive visitors generated revenue from public visitation. A lack of infrastructure and service capacity to collect fees is a significant barrier. Ten pilot public-private partnership agreements are planned in Brazilian national parks with high tourism potential to help expand the revenue-raising potential of protected areas (OECD, 2015).

Tourism fees are an area of opportunity to increase financing for biodiversity in Latin America. For example, Peru charges a fee of USD 10 000 to tourism companies operating in the Manu National Park (Bovarnick et al., 2010).

Administration contracts for protected areas are another way to leverage private financing. Peru has signed at least ten administration contracts with NGOs (sometimes associated with an academic institution) to partially or fully implement management plans on individual protected areas. Contractors often commit to secure and contribute at least the same level of resources as the government. Some have brought in as much as four times the government contribution (IBRD, 2012).

The private sector can provide financial support to supplement scarce public resources. In Brazil, energy companies OGX and MPX committed to support the national parks of Fernando de Noronha and Lençois Maranhenses with more than BRL 4 million each over 2012-18 (Funbio, 2014). In Mexico, an alliance between WWF, SEMARNAT and the Carlos Slim Foundation promised to mobilise USD 100 million to undertake actions that strengthen biodiversity conservation and sustainable development (OECD, 2013).

The private sector can also help finance biodiversity conservation in the form of public-private partnerships. In Brazil, such a partnership was used to support protected area management in what is known as *The Lund Route* – a hiking trail covering 24 km^2 in three protected areas north of the Belo Horizonte metropolitan area. The partnership between a non-profit organisation – Semeia – and the state government of Minas Gerais, aims to increase tourism while improving conservation effectiveness. The initiative began with a bidding process offering a 30-year contract that makes the concessionaire responsible for all conservation activities, including fire control, species control and scientific research (OECD, 2015).

5.4. Biodiversity funds

Biodiversity funds (also referred to as conservation trust funds) can be an efficient tool to finance conservation initiatives, and provide a mechanism for international and private donors to contribute through pooling their resources. They are usually run by private or arms-length institutions entrusted with long-term endowments that support conservation programmes (IBRD, 2012).

The Latin America and the Caribbean Region have 22 conservation trust funds (CTFs) across 15 countries and one transboundary area. The CTFs support 660 protected areas, which include 455 public protected areas and 150 private. The funds invest in protected area equipment and infrastructure, establishment of councils and training, community participation programmes, scientific research and biodiversity monitoring. In general, CTFs have been successful in attracting financing and supporting important biodiversity initiatives. They have also spurred needed capacity building in the region, helping to innovate and share lessons learned (IBRD, 2012). The sources of funds for the CTFs vary, but international donor resources are the most important, followed by private donations, government budget resources and market mechanisms. A number of the funds are endowment funds, where only the interest earned is spent (IBRD, 2012). A significant advantage of CTFs is their independence from government, which provides flexibility and agility in operations and more long term stability in funding. The funds often include members of civil society and the private sector, as well as government officials, on governing boards (IBRD, 2012). However, disadvantages can include high administrative costs, exposure to market volatility, and possible loss of capital. Conditions for successful CTFs include the presence of a long-term fundraising strategy, local ownership over the choice and design of projects supported, widespread stakeholder and government support for biodiversity conservation, a solid legal and financial institutional framework in the country the fund operates in, and clear targets, monitoring and evaluation (Drutschinin and Ockenden, 2015).

The Network of Environmental Funds in Latin America and the Caribbean (RedLAC) was established in 1999 in order to create a system of learning, institutional strengthening, capacity building and co-operation across its 26 members in 16 countries. RedLAC administers a total of USD 328.7 million dedicated to protected areas (IBRD, 2012).

Brazil uses several budget and extra-budgetary funds contribute to financing biodiversity-related expenditure, such as the National Fund for the Environment, Protected Areas Fund, Atlantic Forest Restoration Fund and, most notably, the Amazon Fund – one of the first large-scale efforts to deliver performance-based-payment for greenhouse gas (GHG) emission reductions through forest conservation (Box 5.2). Part of the Amazon Fund is channelled through the Brazilian Biodiversity Fund (Funbio), a non-profit private organisation that raises and invests financial resources for biodiversity conservation, mostly in protected areas, on behalf of the federal and state governments.

The Mexican Fund for the Conservation of Nature was created in 1998, and uses 75% of the interest from its protected areas fund to support innovative and strategic projects implemented by local groups and civil society organisations (IBRD, 2012).

Colombia's Colombia Heritage Fund brings together public and private partners to close the financial gap for the effective management of protected areas and to guarantee the long-term financial sustainability. Under the fund's model, donors mobilise resources for immediate implementation of the required actions while the government commits to gradually increase the allocation of resources and implementation of actions to ensure the sustainability of the system in the long-term.

Box 5.2. The Brazilian Amazon Fund

The Amazon Fund was created in 2008 with the objective to invest in forest conservation and sustainable use, deforestation prevention and monitoring, and to reduce greenhouse gas (GHG) emissions resulting from deforestation and forest degradation.

The Fund was originally based on a performance-based financing mechanism: Norway had committed up provide up to USD 1 billion over a five-year period for bringing GHG emissions from deforestation below a 10-year average. The agreement between Norway and Brazil stipulated that the donation would be made into the Amazon Fund, managed by the Brazilian Development Bank in co-ordination with the Ministry of Environment, and be invested in deforestation control as well as activities to promote the conservation and sustainable use of the Amazon biome.

The fund has a sound monitoring system and has been effective in securing resources for environmental projects, including international and private finance. Between 2009 and 2015, cumulative contributions amounted to USD 970 million, with 72 projects supported. Most funds come from international donors (mainly Norway), but also from companies. At least 80% of the fund's investments are earmarked for the Amazon region and up to 20% can be invested in other Brazilian biomes or tropical countries.

Source: OECD (2015), *OECD Environmental Performance Reviews: Brazil 2015*.

5.5. International financing

International finance is an essential component of biodiversity financing in Latin America, with several international and regional organisations, and bilateral and multilateral agreements, funding conservation initiatives and capacity building throughout the region.

The Global Environment Facility (GEF) is one of the most important funders of biodiversity conservation efforts in Latin America. Biodiversity projects receive more funding from GEF than any other environmental issue in Latin America. Two major GEF-funded projects are the Amazon Region Protected Areas Program (ARPA), with USD 46 million from GEF and USD 121.5 million from other financing, and the Adaptation to the Impact of Rapid Glacier Retreat in the Tropical Andes programme, with USD 9 million from GEF and USD 25 million from other sources (GEF, 2013). The more recent Amazon Sustainable Landscapes Program aims to protect forests, promote sustainable land management and climate change mitigation actions in forest regions in Brazil, Colombia and Peru using USD 5 million from GEF and USD 20 million in co-financing (GEF, 2018).

The Inter-American Development Bank (IDB) began a Biodiversity and Ecosystem Services Programme in 2012 that aims to: integrate the value of biodiversity and ecosystem services into key economic sectors; protect priority regional ecosystems; support effective environmental governance and policy; and create new sustainable development business opportunities (IDB, 2015). For example, the IDB has provided USD 162.5 million for a project in Brazil aimed at the Serra do Mar and Atlantic Mosaics System socio-environmental recovery, including protection of São Paulo water sources (IDB, 2016).

International multilateral and bilateral co-operation and private corporate foundations will continue to be important sources of financing for biodiversity. For example, in the Amazon forest between 2007 and mid-2013, international funding amounted to USD 1.34 billion. Seven of the top donors were from countries and multi-lateral institutions engaged in development co-operation, two were private foundations and one was an international NGO (OECD, 2015). A debt for nature swap between the United States and Brazil has also helped to conserve vulnerable Atlantic coastal rainforest. In 2010, the two countries signed an agreement to convert USD 21 million of Brazilian debt into a fund to protect tropical ecosystems. The money is being used to conserve Atlantic coastal rainforest, as well as the Cerrado and Caatinga ecosystems. While the Atlantic forest once covered most of Brazil's coastline, more than 90% has been cleared. The remaining forest supports significant biodiversity, including 200 bird species and 21 primates endemic to the area (BBC, 2010).

Bilateral official development assistance (ODA) commitments by members of the OECD Development Assistance Committee (DAC) to biodiversity to Latin America and the Caribbean averaged at USD 6.8 billion per year over 2013-15, rising from USD 3.2 billion per year over 2006-07 (at constant 2015 prices). It now represents about 6% of total ODA commitments (OECD, 2018). One of the reasons for the growth is that biodiversity considerations are increasingly being integrated into activities with other primary objectives, and there is growing interest in synergies between biodiversity and climate change actions.

Over 2011-15, seven of the top ten countries with the highest share of biodiversity-related finance in total ODA were located in Latin America and the Caribbean (OECD, 2018). Brazil was the world's top receiving country in this period, with biodiversity-related ODA commitments reaching USD 270 per year (at constant 2015 prices); equal to more than a quarter of total ODA (OECD, 2018). ODA where biodiversity is a secondary, but significant, focus has grown and is now greater than ODA targeting biodiversity as a primary objective.

Figure 5.2. Biodiversity-related ODA commitments are significant

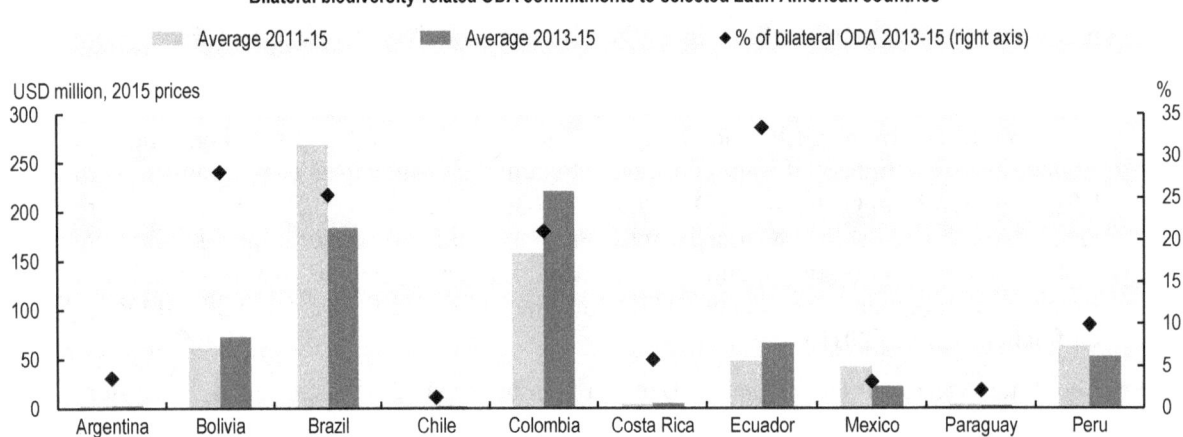

Note: For many countries less than half of the total ODA is screened against the Rio marker for biodiversity.
Source: OECD (2018), "Creditor Reporting System: Aid activities targeting Global Environmental Objectives", OECD *International Development Statistics* (database).

StatLink https://doi.org/10.1787/888933886132

References

BBC (2010), *US-Brazil Debt for Nature Swap to Protect Forests*, BBC News, 12 August 2010, www.bbc.com/news/world-latin-america-10958695.

Biofin (2016), *Biofin: The Biodiversity Finance Initiative Brochure*, United Nations Development Programme, New York, http://biodiversityfinance.net/.

Bovarnick, A., J. Fernandez Baca, J. Galindo, and H. Negret (2010), *Financial Sustainability of Protected areas in Latin America and the Caribbean: Investment Policy Guidance*, United Nations Development Programme (UNDP) and The Nature Conservancy (TNC), New York and Arlington, www.cbd.int/financial/finplanning/g-planscorelatin-undp.pdf.

CBD High-Level Panel (2014), Resourcing the Aichi Biodiversity Targets: An Assessment of Benefits, Investments and Resource needs for Implementing the Strategic Plan for Biodiversity 2011-2020. Second Report of the High-Level Panel on Global Assessment of Resources for Implementing the Strategic Plan for Biodiversity 2011-2020. Montreal, www.cbd.int/financial/hlp/doc/hlp-02-report-en.pdf.

Drutschinin, A. and S. Ockenden (2015), *Financing for Development in Support of Biodiversity and Ecosystem Services*, OECD Development Co-operation Working Papers, No. 23, OECD Publishing, Paris. http://dx.doi.org/1787/5js1sqkvts0v-en.

Funbio (2014), Funbio and Protected Areas, 2014, Brazilian Biodiversity Fund, Rio de Janeiro, www.funbio.org.br.

GEF (2018), Amazon Sustainable Landscapes Program, information sheet, Global Environment Facility, https://www.thegef.org/sites/default/files/publications/Amazon-Sustainable-Landscapes-Program-2018.pdf.

GEF (2013), *GEF and Latin America and the Caribbean: Map*, Global Environment Facility, http://beta.thegef.org/sites/default/files/documents/gef-LAC-Map-June_7E_-_CRA_0.pdf.

IBRD (2012), *Expanding Financing for Biodiversity Conservation: Experiences from Latin America and the Caribbean*, International Bank for Reconstruction and Development, World Bank, Washington D.C, www.worldbank.org/content/dam/Worldbank/document/LAC-Biodiversity-Finance.pdf.

IDB (2016), *Serra do Mar and Atlantic Forest Mosaics System Socioenvironmental Recovery Project*, Inter-American Development Bank, Washington D.C., www.iadb.org/en/projects/project-description-title,1303.html?id=BR-L1241.

IDB (2015), *Biodiversity and Ecosystem Services Program: An Overview*, Inter-American Development Bank, Washington D.C., http://idbdocs.iadb.org/wsdocs/getdocument.aspx?docnum=38186826.

OECD DAC (2010), *Policy Statement on Integrating Biodiversity and Associated Ecosystem Services into Development Co-operation*, Development Assistance Committee, www.oecd.org/environment/environment-development/46024461.pdf.

OECD (2018), "Creditor Reporting System: Aid activities targeting Global Environmental Objectives", *OECD International Development Statistics* (database), http://dx.doi.org/10.1787/9c778247-en (accessed on 26 January 2018).

OECD (2015), *OECD Environmental Performance Reviews: Brazil 2015*, OECD Publishing, Paris, http://dx.doi.org/10.1787/9789264240094-en.

OECD (2013), *OECD Environmental Performance Reviews: Mexico 2013*, OECD Publishing, Paris, http://dx.doi.org/10.1787/9789264180109-en.

OECD DAC (2010), *Policy Statement on Integrating Biodiversity and Associated Ecosystem Services into Development Co-operation*, Development Assistance Committee, www.oecd.org/environment/environment-development/46024461.pdf.

OECD/ECLAC (2016), *OECD Environmental Performance Reviews: Chile 2016*, OECD Publishing, Paris, http://dx.doi.org/10.1787/9789264252615-en.

6. Mainstreaming

This chapter discusses the importance of aligning sectoral and other policies with biodiversity objectives. It focusses on the sectors that pose particular pressures on Latin American biodiversity, such as agriculture, fisheries, forestry, energy and infrastructure development, and tourism. The final section highlights synergies in policy approaches that benefit both biodiversity and climate change goals.

The statistical data for Israel are supplied by and under the responsibility of the relevant Israeli authorities. The use of such data by the OECD is without prejudice to the status of the Golan Heights, East Jerusalem and Israeli settlements in the West Bank under the terms of international law.

6.1. Introduction

Latin America's rapid population and economic growth is creating opportunities, and challenges, for biodiversity conservation and sustainable use. Mainstreaming and aligning sectoral and other policies with biodiversity objectives will become increasingly important in the region, as development continues and most areas remain outside official protection. The region's population (including the Caribbean) is expected to reach 700 million by 2030, with 500 million categorised as middle-class, and GDP is expected to double from current levels (IDB, 2015). GDP growth is creating opportunities to reduce poverty, but is also increasing pressures on biodiversity, as people's consumption habits change and production adapts accordingly. International calls for mainstreaming biodiversity into development have grown over the past decade, such as through the CBD and the Sustainable Development Goals (in particular Goals 14 and 15), and the OECD Development Assistance Committee's Policy Statement on Integrating Biodiversity and Associated Ecosystem Services into Development Co-operation (OECD DAC, 2010).

While there have been improvements in the integration of biodiversity considerations into sectoral policies in Latin America, significant further work is required to ensure that the approaches are comprehensive, consistent, effective and accepted by local communities. Mainstreaming is particularly important for policies and programmes related to agriculture, fishing and aquaculture, forestry, tourism, mining, energy and infrastructure development because these sectors are heavily dependent on natural resources and the services that healthy, well-functioning ecosystems provide, but are also sectors whose activities can have substantial negative effects on biodiversity. Health, processing and manufacturing and construction are also recognised by the CBD as priority sectors for mainstreaming. There are significant benefits to be gained from aligning biodiversity considerations with climate change mitigation and adaptation policies and programmes.

There are many opportunities for governments to mainstream biodiversity considerations at sector level. For example, sectoral strategies, action plans and programmes, industry standards, sector guidelines and good practices, and certification schemes offer potential for adjustment (Van Winkle, 2015). It is also important to mainstream development considerations into biodiversity policies and to mainstream biodiversity into development co-operation portfolios (Drutschinin et al., 2015). Mainstreaming is preferable to isolated policy development in that it allows for integrated approaches that consider economic, environmental and social objectives, allowing for greater potential to achieve optimal outcomes that can be sustained into the future.

Mainstreaming biodiversity is not something that can be done overnight. It is a complex process that requires sustained investment and engagement over long time periods (at least 10-15 years), relationship-building, high-level buy-in and managing trade-offs. Enabling conditions are also key factors for success. Better, higher-resolution and accessible data and analysis on the status, trends and value of ecosystems and species can support mainstreaming efforts by helping to establish environmental baselines in areas where development is occurring, and identifying protection priorities. Good governance, effective processes and strong institutions are essential. In addition, effective partnerships and open dialogue with external stakeholders can help improve engagement, support and the business case for biodiversity (Drutschinin et al., 2015; Huntley et al, 2014).

Development co-operation providers can facilitate the biodiversity mainstreaming process in developing countries, such as in those in Latin America, through both financial and technical assistance. This can include assistance with national and sectoral plans, but also with data gathering, assessment tools (e.g. ecosystem valuation and cost-benefit analysis), and the design and implementation of informational, regulatory and economic instruments that support biodiversity mainstreaming (Drutschinin et al., 2015).

6.2. Agriculture

Agriculture is a significant source of income and employment in Latin America. With roughly 8% annual growth in exports since the mid-90s, agricultural products now make up around 25% of total exports. Latin America represents 13% of agricultural trade (World Bank, 2013). In Peru, close to half of workers are employed in agriculture or retail and restaurants (OECD, 2015a). In Chile, agricultural production grew by 27% between 2002 and 2013, and the country has become one of the world's leading exporters of fresh fruit and wine (OECD/ECLAC, 2016).

However, Latin America's growing agriculture sector is also putting increasing pressure on fragile terrestrial and aquatic ecosystems as a result of land-use change and deforestation, livestock grazing and effluents, water use, and pesticides and fertilisers (Chapter 2). The challenge is to find ways to grow productivity rather than agricultural area, promote sustainable cultivation and irrigation practices, and protect areas with high biodiversity value. While trade agreements and market demand are influencing a trend towards more sustainable production practices and organic products, sustainable practices continue to lag in most countries and the proportion of organic agricultural production remains small.

Pesticide and fertiliser use is a major source of water and soil pollution. Most Latin American countries are increasing their fertiliser consumption, albeit it remains modest compared to many OECD countries. Most Latin American countries have also increased their pesticide use, though Colombia managed to reduce theirs between 2006 and 2011 (Figure 6.1). While comparable data was not available for Brazil, the country's agriculture sector is considered one of the world's top consumers of fertilisers and pesticides (ABRASCO, 2015). High pesticide sales may partly be explained by the fact that Brazilian agriculture is practiced in tropical and subtropical environments, with a high incidence of pests. In Chile, rapid growth in the use of pesticides per unit of agricultural land since 2000 has been linked to the death of bee populations (CIAP, 2012).

Figure 6.1. Most Latin American countries are increasing their pesticide use

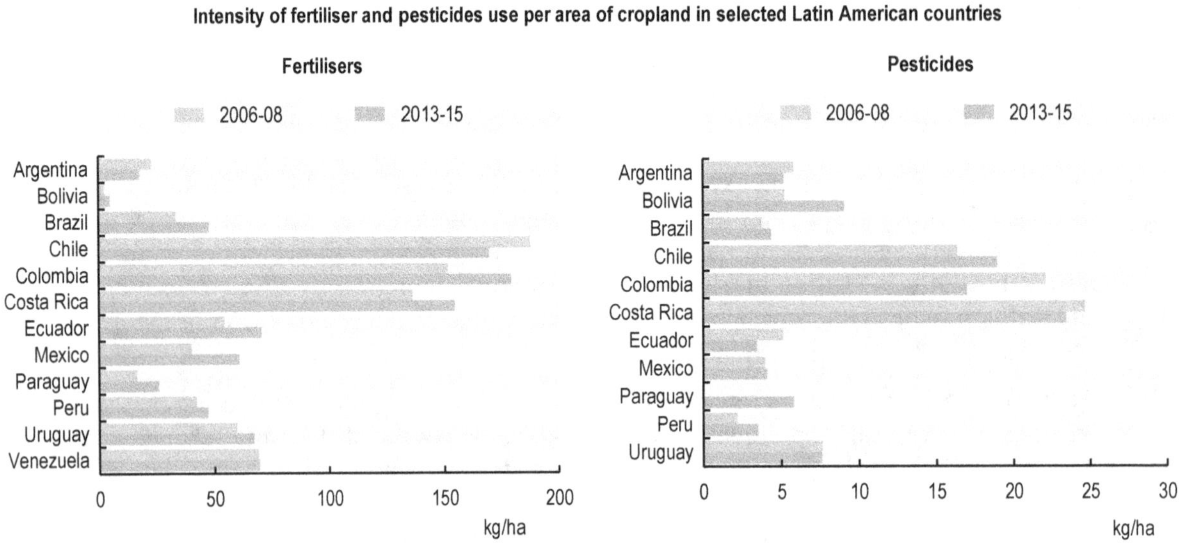

Intensity of fertiliser and pesticides use per area of cropland in selected Latin American countries

[Bar chart: Fertilisers (kg/ha, 0–200) and Pesticides (kg/ha, 0–30) for Argentina, Bolivia, Brazil, Chile, Colombia, Costa Rica, Ecuador, Mexico, Paraguay, Peru, Uruguay, Venezuela; comparing 2006-08 and 2013-15.]

Note: Definitions may vary considerably among countries and limit international comparability.
Fertilisers: data for Colombia are partial and cover less than 70% of the country. Pesticides: while data generally refers to sales of pesticides expressed as active ingredients, for some countries data refer to imports and are expressed formulated products. Brazil: partial data and different from national sources: Costa Rica: latest data refer to 2011. Fertilisers: data covering less than 70% of the country.
Source: FAO (2018), *Agro-environmental Indicators* (database), http://faostat3.fao.org/download/E/EP/E.
StatLink https://doi.org/10.1787/888933886151

Many Latin American countries are struggling to integrate biodiversity considerations and efficient natural resource use into the agriculture sector. Agricultural support systems have not yet been reformed to discourage pesticide use (Section 4.3); Mexico, for example, still provides VAT exemptions for agrochemicals (OECD, 2013), and pesticides and fertilisers are exempt from some federal and state taxes in Brazil (OECD, 2015b). While Peru has improved its institutions and instruments for environmental management in agriculture, challenges remain in terms of the widespread lack of land title, slash-and-burn techniques for converting forest to farmland, and an overall unplanned approach to agriculture (OECD/ECLAC, 2017). Colombia has an ambitious livestock strategy that calls for the return of 10 million ha of pasture to a more natural state through reforestation or conversion to silvopasture[1] and the intensification of cattle rearing and agriculture. Progress was also made with regard to closure of the agricultural border, reforms in land access and use, formalisation of property, and protection of reserve areas as part of the commitments of the Peace Agreements. However, Colombia will be challenged to achieve its goals without reforming incentives that promote the expansion of grazing land – such as property tax exemptions and agricultural credits (OECD/ECLAC, 2014).

Agriculture is the largest water user in most regions (UN Water, 2014), and represents over 80% of total water withdrawal in Chile and Peru. This contrasts with the OECD average, where the greatest water withdrawal is from industrial sectors (Figure 6.2). In Chile, water demand is increasingly exceeding supply in the central parts of the country, where agricultural production is concentrated. The Chilean agriculture sector's water demand further threatens biodiversity by draining wetlands and eroding soil (OECD/ECLAC, 2016). In Peru, export agriculture is heavily practiced in water-deficit zones that cannot guarantee its sustainability (OECD/ECLAC, 2017). Water scarcity reduces the ability of water bodies to eliminate excessive nutrients, thereby contributing to eutrophication.

Figure 6.2. Agriculture is the largest water user in most countries

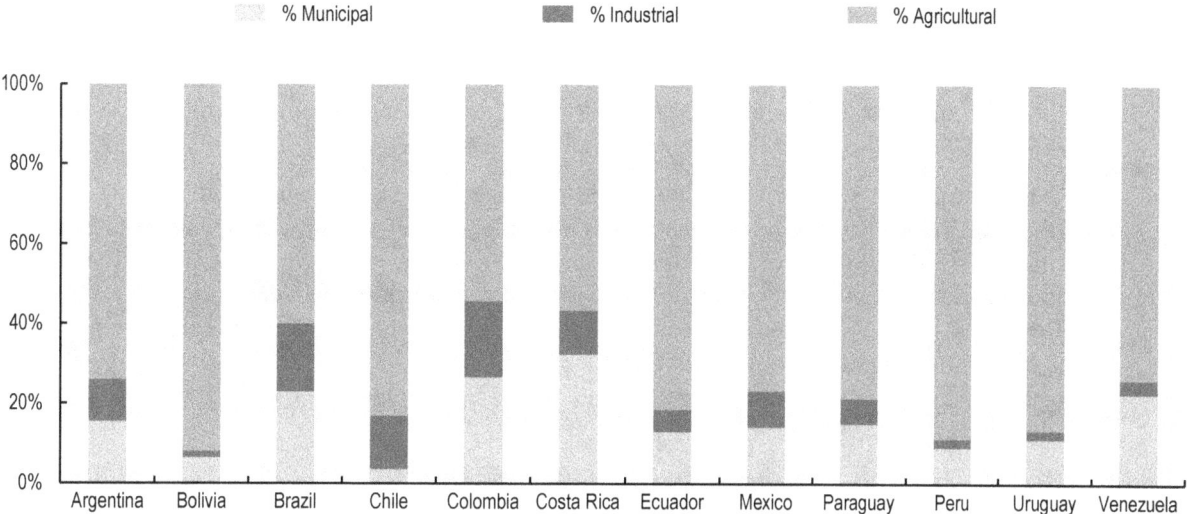

Note: Data is for the latest year available, ranging from 2000 to 2013.
Source: UN Water (2014), Key Water Indicators Portal, www.unwater.org/kwip.
StatLink https://doi.org/10.1787/888933886170

Despite growing water scarcity risks, irrigation practices have yet to significantly shift to modern water saving practices (Box 6.1). Irrigation subsidies have encouraged the adoption of modern water-saving techniques, but older methods still account for 70% of irrigated area and Chile has among the highest irrigation water application rates in the OECD (OECD/ECLAC, 2016). Mexico has a programme that provides financial incentives for water conservation, but farmers' uptake of the payments has been limited and subsidy programmes continue to support irrigation, particularly for electricity used to pump water (OECD, 2013). In Brazil, water abstraction is not charged in many regions (OECD, 2015b).

> **Box 6.1. Irrigation practices in Latin America**
>
> Agricultural irrigation is responsible for the largest proportion of water use in Latin America, with expectations for significant growth to support export and feed growing domestic populations. At the same time, many arid and semi-arid regions are experiencing water shortages, exacerbated by climate change, that threaten biodiversity and create conflict.
>
> Shifting agricultural practices and technologies to become more water efficient is possible through design and modernisation of irrigation approaches. Improved knowledge of crop water requirements is also important.
>
> Approximately 95% of irrigated lands in South America are surface irrigated, with the remainder using more water-efficient sprinklers or drip and micro-sprinkler irrigation, highlighting the potential to improve water productivity. Brazil has managed to achieve 35% sprinkler irrigation and 6% drip and micro-sprinkler irrigation. New methods and tools for irrigation include deficit irrigation (optimising irrigation to apply water during drought-sensitive growth stages of a crop) and remote sensing (using satellites to obtain regular water management information feedback from the field).
>
> *Source*: de Oliveira et al. (2009), *Irrigation Water Management in Latin America*.

Policies for the protection and conservation of agro-biodiversity in Peru have been strengthened but resources remain inadequate. While policy initiatives are in place, they have not borne fruit, and there is limited support to develop germplasm banks of native crops or research on native and introduced species. Peru is considered one of the "centres of origin" for farming in the Americas, and the Andean zone has the second largest variety of maize after Mexico. Peru has a wealth of native plants that could be strategically used for climate change adaptation, given that they are particularly efficient in their use of water (OECD/ECLAC, 2017).

Brazil has begun to make progress in greening agricultural support systems, by making access to subsidised rural credit in the Amazon biome conditional on the legitimacy of land claims and compliance with environmental regulations. Starting in 2017, rural credit will also be conditional on land registration in the Rural Environmental Cadastre. Other subsidies include the Family Production Socio-economic Development Programme which awards farmers and ranchers with up to one-third of the minimum wage when they use more environmentally sound production practices, and the Low-Carbon Agriculture programme that provides subsidised credits for implementing good environmental practices. However, the volume of programmes supporting sustainable agriculture is small compared to the total support provided to farmers (OECD, 2015b).

6.3. Fishing and aquaculture

Fishing and aquaculture are important sources of employment and income in Latin America. Peru and Chile have some of the largest fisheries in the world, and production in Mexico and Brazil is also significant. Peru has a major industrial-scale maritime fishery, with anchovies accounting for 86% of the catch, and is also the world's leading producer of fishmeal and fish oil. Fish catches are generally declining in the region, while production from aquaculture is increasing (Figure 6.3). For example, in Chile, fish catches halved between 2005 and 2010, due to low fish stocks and overexploitation, but aquaculture production has increased to make the country one of the world's largest producers. Aquaculture production in Brazil grew five-fold between 2000 and 2015. Fishing and aquaculture threaten marine and aquatic biodiversity due to overfishing, bycatch, invasive species, disease and pollution (FAO, 2011) (Chapter 2). Further effort is needed to improve monitoring of marine and aquatic ecosystems, and develop and enforce more sustainable fishery and aquaculture policies.

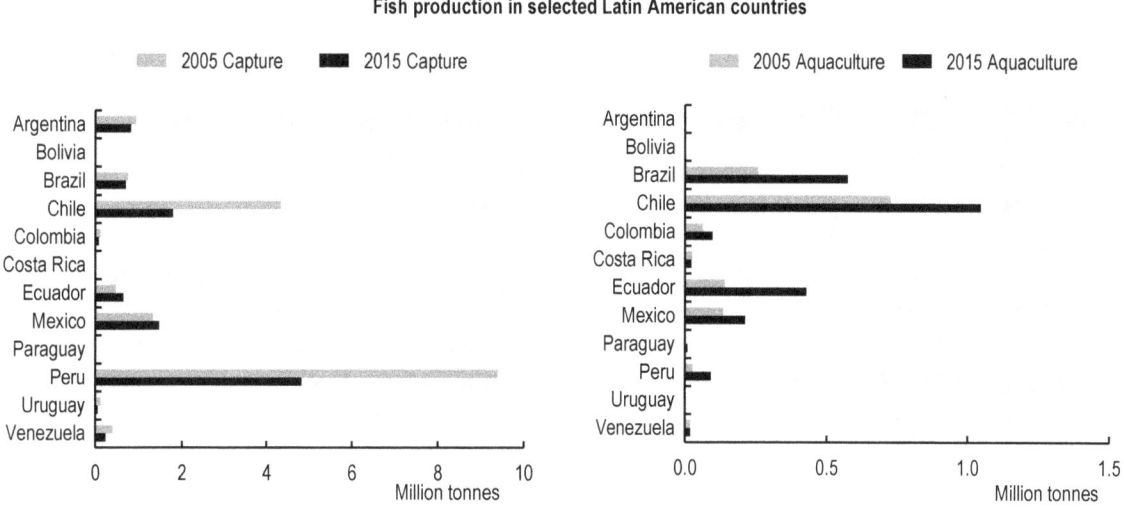

Figure 6.3. Fish catches are declining, while aquaculture is growing

Fish production in selected Latin American countries

Note: Data exclude marine mammals, miscellaneous aquatic products and aquatic plants.
Source: FAO (2018), Fisheries and Aquaculture Statistics, http://www.fao.org/fishery/statistics/en.
StatLink https://doi.org/10.1787/888933886189

A number of regulatory, economic and information instruments for fisheries and aquaculture management have been introduced and strengthened in Latin America over the past decade. In Peru, improvements have been made to fishing season limits, quotas, and minimum size, with the anchovy quota shifted from aggregate to per vessel, helping to reduce the size of the fleet and the number of processing facilities (OECD/ECLAC, 2017). Chile has embraced a quota system to manage its fisheries, and has a transferable quota licence system for industrial fisheries covering a part of the industrial sector's overall quota (Section 3.3). In 2013, the country's Law on Fishing and Aquaculture was amended to shift the basis for quota establishment from economic and social considerations to scientific and technical factors. The law also introduced concepts such as the precautionary principle and ecosystem approaches, and reserves the first nautical mile from shore exclusively for smaller vessels. In 2014, Chile implemented a new tax on Fisheries Law Extraction Rights based on the quota size of each industrial operator. The certification of salmon production centres to best practices in the country has increased (OECD/ECLAC, 2016).

Despite progress in the use of a variety of instruments for fishery and aquaculture management, governance is fragmented and monitoring insufficient. In Peru, for example, responsibility for the ocean is divided among many agencies, with co-ordination through the Multisectoral Commission for Environmental Management of the Coastal Marine Environment (COMUMA). While industrial fisheries are subject to remote tracking through a satellite monitoring system, a significant portion of marine and inland fisheries and aquaculture activities have little to no supervision as a result of limited human and financial resources dedicated to monitoring and enforcement (OECD/ECLAC, 2017). In Chile, while the Fisheries Act is being amended to limit emissions of solid and liquid waste from aquaculture, constrained resources for monitoring and enforcement is slowing progress (OECD/ECLAC, 2016). Brazil's shared fishery management model is challenged by insufficient mechanisms to monitor and control compliance, and difficult co-operation between the Ministry of Environment and the Ministry of Fisheries and Aquaculture (MMA, 2015).

Effective fishery management requires good data on the status and trends of species and ecosystems to identify priorities for action. For example, while Colombia uses catch quotas established by the Ministry of Agriculture with scientific support from the National Aquaculture and Fisheries Authority (AUNAP) and the executive committee on Fisheries, the lack of data on commercial fish species has posed a significant challenge to developing effective biodiversity policy (OECD/ECLAC, 2014). Limitations in data on aquatic habitats and fishery resources are is also a challenge in Brazil (OECD, 2015b).

Local initiatives targeting fishing communities and artisanal fishers are proving to be an important component of biodiversity conservation and sustainable use in Latin America. Chile has established over 700 Areas of Management and Exploitation of Benthic (bottom-dwelling) Resources where exclusive rights are assigned to organisations of artisanal fishers. The North of Choco Department in Colombia has implemented co-operative management initiatives where local fishery communities are involved in the development and implementation of sustainable fishery policy. These approaches could be replicated in other coastal areas.

Brazil's fishery and aquaculture management is in dire need of improvement. Currently, most of the country's fisheries involve obsolete fleets targeting overexploited fish stocks. Brazil has no formal environmental licencing required for fishing activities though there are restrictions on fishing periods, areas and gear, and aquaculture activities are subject to licensing. The 2015 OECD EPR noted the need for additional measures, including fish catch quotas, more effective management plans for overexploited species, and extension of marine protected areas, particularly in areas where fish stocks are at their limit (OECD, 2015b).

6.4. Forestry

Forestry is an important economic sector in some Latin American countries, accounting for 5.2% and 7.3% of Chile and Brazil's exports respectively (OECD, 2015b; OECD/ECLAC, 2016). In all Latin American countries, both forestry and deforestation practices pose a major threat to biodiversity (Chapter 2). While commercial forestry practices have improved and efforts to preserve and expand native forests have been strengthened, further efforts are needed to address illegal deforestation (Section 2.2) and direct consumption (FAO, 2015).

Most major producers of forest products in Latin America increased their use of forest certification between 2005 and 2014. However, certified forest still represents a relatively small proportion of forest area designated for production (Figure 6.4) (FAO, 2015). Brazil, Chile and Peru have improved certification rates in their forestry sectors. For example, Chile now has at least 70% of plantation companies affiliated with the trade association qualified for Forest Stewardship Council (FSC) certification (OECD/ECLAC, 2016).

Mexico has two forest certification schemes; the Mexican Standard for the Certification of Sustainable Forest Management, which makes products eligible for green public procurement, and the FSC label (OECD, 2013). Brazil has two national certification schemes as well as the FSC (OECD, 2015b).

Figure 6.4. **Certified forest still represents a relatively small proportion of total forest area**

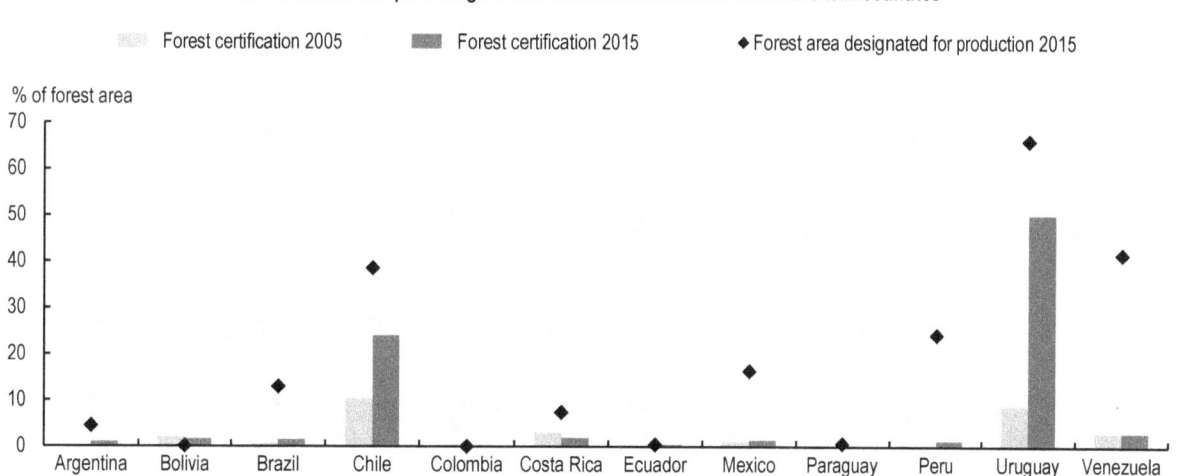

Note: Most forest certification is through the FSC (Forest Stewardship Council), though Brazil and Chile also use PEFC (Programme for the Endorsement of Forest Certification), and Mexico has two domestic certification programmes.
Source: FAO (2015) *Global Forest Resources Assessment 2015*, www.fao.org/forest-resources-assessment/en/.
StatLink ⛓ https://doi.org/10.1787/888933886208

Some countries implement measures beyond certification schemes in order to promote sustainable forest management. For example, Brazil's 2006 Forests Management Law introduced concessions as an instrument to promote sustainable forest management for timber production. Under the law, federal, state and municipal governments can grant the legal right for private companies to harvest timber and non-timber forest products, provided that the forest is sustainably managed. However, the concessions have been slow to expand due to a lack of expertise both in companies and government, high concession fees and unsolved land tenure conflicts. Additionally, a large number of rural land holdings do not comply with forest conservation obligations set in the 2012 Forest Code (OECD, 2015b). Chile's 2008 Native Forest Recovery and Forestry Promotion Law created a financial incentive for the protection and preservation of native forests, and a Conservation Fund to promote management, conservation, restoration and research on native forest ecosystems (OECD/ECLAC, 2016).

Many Latin American countries provide incentives to promote reforestation. Brazil's National Plan for Native Vegetation and Recovery (Planaveg) aims to promote large-scale forest restoration, targeting 125 000 km^2 within 20 years. It intends to do this by raising awareness, making seedlings available and affordable, creating markets for products from restored forests and introducing new finance mechanisms, *inter alia* (OECD, 2015b). The Mexican government runs a national reforestation programme, PRONARE, which gives support to landowners/users for reforesting degraded forest land, providing seedlings, training and funding. Since 2007, 1.87 million ha have been reforested (OECD, 2013). In both countries, high reforestation costs are a barrier to further progress. To maximise the impact of funds, they should be targeted at priority areas, e.g. those that are most important for biodiversity protection and ecosystem service provision (OECD, 2013; OECD, 2015b).

Increased forest area does not necessarily mean positive trends for biodiversity. For example, Chile's forest expansion, encouraged by long-standing afforestation and forest plantation subsidies, has primarily consisted of non-native tree species plantations, such as Radiata Pine and Eucalyptus. While these plantations have climate change and soil erosion benefits, they can also increase pressure on native vegetation, habitat-specific species and water resources (OECD/ECLAC, 2016). Colombia encourages the planting of native species by providing larger subsidies for native planting than introduced species. The government subsidises 50% of the up-front planting costs for certified introduced species and 75% for certified native species through its Forestry Incentive Certification Programme. Between 1995 and 2011, the initiative supported reforestation of 173 950 ha. The initiative will, however, need additional funding and stronger monitoring, reporting and verification to achieve the government's goal of reforesting one million ha with 60% commercial plantation (OECD/ECLAC, 2014).

Box 6.2. Slowing deforestation in Brazil: Progress and challenges

Brazil has the second largest forest area in the world and is home to the world's largest rainforest. Two-thirds of the country is covered with forest or other wooded land. The Amazon represents 30% of the world's tropical forest, hosting 600 types of terrestrial and freshwater habitats and the Cerrado region is one of the world's biodiversity hotspots.

The rate of deforestation has slowed significantly in Brazil. However, Brazil has the largest number of hectares of forest lost each year in the world. Forest area has decreased by 5% between 2000 and 2015, and by 10% when compared to 1990.

Unclear legal tenure, especially in the Amazon, has been a major driver of deforestation. In 2011, only 4% of the Amazon area had a valid private property title. In 2006, the government pledged to reduce deforestation in the Amazon by 80% by 2020. Annual deforestation in the region dropped from 27 700 km^2 in 2004 to 4 800 km^2 in 2014. Brazil's NDC pledges to end illegal deforestation in the Amazon by 2030.

Brazil has innovative instruments helping to achieve these reductions in deforestation. For example, Brazil's Action Plan for the Prevention and Control of Deforestation in Amazonia Legal (PPCDAm), which aims to clarify land tenure, strengthen monitoring, enforcement and compliance, and promote sustainable production chains that provide alternatives to deforestation, has been held up as a model for other countries. A new Forest Code and its innovative implementation and enforcement instruments promise to help further reduce illegal forest clearing and reconcile the objectives of agricultural development and biodiversity conservation. Effective implementation will depend on having sufficient financial and human resources and improving co-ordination across levels of government.

Source: FAO (2015), *Global Forest Resources Assessment 2015*; OECD (2015), *OECD Environmental Performance Reviews: Brazil*.

Mining is an important part of many Latin American economies, and Latin American countries are some of the world's largest mineral producers. Chile and Peru are the world's first and third largest copper producers respectively, and Mexico and Peru are the first and third largest silver producers. Brazil is the third largest producer of iron ore (USGS, 2016a, 2016b, 2015a, 2015b). Ore and metal exports represent roughly half of merchandise exports for Chile and Peru, and over 10% for Brazil (Figure 6.5). The oil and mining sectors represent more than half of Colombia's exports (OECD/ECLAC, 2014). However, mining is also a major driver of ecosystem degradation (Chapter 2). Mining and illegal mining are also an important source of social conflicts that hinder a proper management and governance of biodiversity; for example 20 of the 30 cases of environmental conflict documented in Chile are linked to mining activities (Segall, 2014).

Figure 6.5. Mining is an important part of many Latin American economies

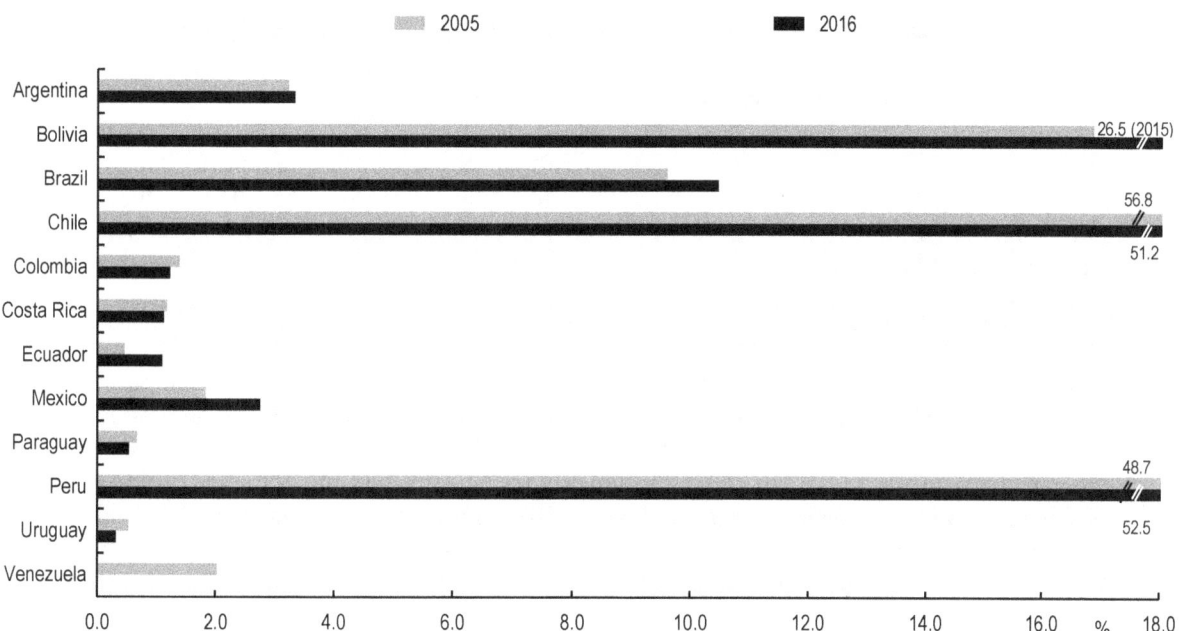

Source: World Bank (2018), World Development Indicators (database), http://data.worldbank.org/data-catalog/world-development-indicators.

StatLink https://doi.org/10.1787/888933886227

Important steps are being made to improve environmental performance in the mining sector. For example, Environmental Impact Assessments (EIA) of large mining projects have improved, and the use of tools such as biodiversity offsets, where biodiversity loss at the site is compensated with conservation projects in other locations, has increased with examples in Peru and Chile (OECD/ECLAC, 2016) (Section 3.3). However, more effort is needed on enforcement and monitoring, effective Environmental Impact Assessment processes (biodiversity considerations are not always consistently and comprehensively addressed), accelerated clean-up of abandoned mines, and avoiding conflict with indigenous and local communities.

Several incidents have highlighted the significant risks associated with mining projects that do not have adequate oversight. Tailings ponds of hazardous mining waste present a risk to humans and ecosystems, particularly in regions prone to earthquakes, landslides and heavy rains. The collapse of a mining dam in the Brazilian state of Minas Gerais in

November 2015, for example, where communities below the mountain mine were covered with toxic mud, caused both human casualties and environmental damage from the flooding of the Rio Doce river with mining waste that killed aquatic species and impacted the source of drinking water for thousands of people (Phillips, 2015). Since the tragedy, there have been calls for improved regulations and enforcement at Brazilian mines. Many residents in the flooded town were not aware of the risks above, generating mistrust (Phillips, 2015). Minimising the risks of resource-based mining development requires transparent and robust legal frameworks and fiscal regimes that are implemented and monitored by strong governmental and societal institutions (Lizanco Rodriguez et al., 2013).

Small-scale and artisanal mining also represents a risk to biodiversity as it is often poorly monitored or controlled. In Peru, artisanal miners (mainly of gold) often operate without any environmental permit, increasing the likelihood of pollutant releases into water and soil. Peru's government is, however, actively seeking to formalise small-scale and artisanal mining, eradicate illegal mining and improve environmental performance. Authorities now have the legal right to conduct environmental audits of mining activities of this nature, which are pursued without any operating or environmental permit, and The Corrective Environmental Management Instrument applies to existing small-scale and artisanal mining operations that are in the process of formalisation, to bring them into line with legally determined environmental obligations. In Chile, where the regulatory focus has been on large-scale mines, there is insufficient information available on the environmental impacts of small-scale mining operations, which are subsidised by the government (OECD/ECLAC, 2016).

The rapid expansion of mining development, particularly in rural, poor areas and on or adjacent to lands occupied by indigenous peoples, has led to growing conflict in numerous Latin American countries. Many of the concerns relate to the impact of mining activities on land and water. Peru has been a leader in tackling social conflict and improving transparency in the sector. In 2012, the government established the National Office for Dialogue and Sustainability (ONDS) to help resolve mining-related disputes. Peru was also the first country in Latin America to successfully implement the accountability standard of the Extractive Industries Transparency Initiative (EITI), which aims to improve the transparency of tax revenues and payments from mining and the extent to which they flow back into the development of mining areas (OECD/ECLAC, 2017).

Strained dialogue and co-operation between different parts of government can be a barrier to the effective enforcement of environmental policies in the mining sector. In Colombia, environmental authorities have been unable to prevent the Ministry of Mines and Energy from granting mine titles in areas of environmental importance over the past decade. In 2010 alone, over 400 mining titles were granted in protected areas, and over 1 000 in wetland habitats and 2 000 in forest reserves. In 2013, however, the Ministry of Mines and Energy and the Environment Ministry signed an agreement to secure protected areas from development and pursue sustainable development within the sector. The mining ministry has also established an office to deal with social and environmental issues, and the two ministries are conducting research on the impact of mining on natural resources (OECD/ECLAC, 2014).

Several countries have taken steps to limit future environmental damage from companies currently operating mines. However, clean-up of historical damage remains limited due to a lack of legal frameworks making companies liable to do this. Abandoned mines represent a significant ongoing risk to soil and water contamination in Latin America. Peruvian legislation on the treatment and clean-up of environmental mining liabilities (PAMs) could provide a model for the region. The first step taken was to draw up an inventory of abandoned sites, with 8 616 PAMS as of 2015, 50% of which were determined to pose a high or very high risk. The organisation Activos Mineros pursues remediation at sites abandoned by former state-owned mining enterprises. Currently-operating mining companies in Peru are also liable for the closure of mines, and are required to take measures to avoid risks to human health and the environment from abandoned mines (OECD/ECLAC, 2017). While Chile has made progress in identifying abandoned or inactive mine sites, and a new 2012 law requires all new mines to have approved end-of-life closure plans (Box 6.3), there are no decontamination plans in place for its estimated 650 abandoned mining sites (OECD/ECLAC, 2016).

Box 6.3. Chile's mine closure financial guarantees

Chile's 2012 Mine Closure Law aims to prevent the creation of abandoned mine sites in the future by requiring mining companies to provide financial guarantees for each operation and develop detailed mine closure plans. This mechanism is meant to generate sufficient funds for site closure should the operator default on its decommissioning obligations. The first phase of the law's implementation affected every mine over a minimum size threshold with an approved closure plan. These companies had to provide, by November 2014, a cost estimate that took into account remaining mine life and a discount rate based on a state-provided index. Once the estimate was approved, the mining company had to provide a guarantee for the amount, using one of the approved financial instruments.

Initially, 20% of the present value is required to be guaranteed. The amount gradually increases over 15 years (or two-thirds of the remaining mine life, whichever is shorter) to the full present value of closure costs. The law allows for partial reductions of the guarantee. The total amount to be guaranteed is estimated to be USD 30 billion.

Source: OECD/ECLAC (2016), *OECD Environmental Performance Reviews: Chile 2016*; Weeks (2015), *Mine closure in Chile – challenges and changes*.

6.5. Energy and infrastructure

Latin America produces energy products for both export and domestic consumption, making it an economically important sector in the region. While the region has relatively large oil reserves, they are concentrated in few countries. Mexico, Brazil and Colombia are major oil producers and exporters (IEA, 2016), and oil is also a factor in Peru's economy. Natural gas and coal reserves are not as significant in Latin America (IEA, 2015). The region's electricity consumption is projected to grow by 75% between 2009 and 2035 (Tissot, 2012). While electricity coverage has increased substantially, there remain significant populations without coverage in rural areas, particularly in countries such as Peru. Many sources of generation will be needed to meet this demand – hydropower, wind power, oil, natural gas, coal and biomass (largely from sugarcane residues), each of which pose their own risks and challenges as far as biodiversity is concerned. The expansion of road and industrial infrastructure is another major driver of land-use change and habitat loss and fragmentation in Latin America (Chapter 2).

Hydroelectric development continues to be contentious in many regions of Latin America. While it is an important source of energy to limit greenhouse gas (GHG) emissions and air pollutants, the projects can also result in displacement of people and destruction of natural habitat for the creation of reservoirs. Hydropower dominates electricity generation in Brazil and is also an important electricity source in Chile, Colombia and Peru (OECD/ECLAC, 2016; OECD, 2015b; OECD/ECLAC, 2014). Across Latin America there are more than one thousand dams measuring 15 metres tall or more (Cevallos, 2006), and this number is expected to grow as governments seek ways to meet growing energy demands at the same time as climate change commitments. Carefully managing hydroelectric expansion, and using the latest technologies and approaches to limit ecosystem impacts, will be important in Latin America in the coming years.

To take an example, in Brazil, hydropower represents almost three-quarters of electricity production. While this share has declined over the past decade, there remains substantial growth potential to meet rising demand. However, most potential is located in the Amazon, creating challenges for environmental licensing and public acceptability. While hydropower projects are subject to environmental licensing and impact assessments, unlike other countries Brazil has not paid significant attention to the impact on water flows needed to sustain freshwater ecosystems and ecosystem services. Impacts have also generally been addressed as mitigation measures late in the process rather than early in the planning stages. The 2015 OECD EPR suggested using strategic environmental assessments to identify where energy capacity could be built with the least environmental impact, taking into consideration cumulative impacts (OECD, 2015b).

Strategic environmental assessment (SEA) and environmental impact assessment (EIA) are key tools to integrate biodiversity considerations into energy and other infrastructure plans and projects. While EIA has existed in most Latin American countries for some time and SEA is increasingly being used, the design and implementation of both tools needs to improve significantly to consistently and comprehensively take biodiversity into consideration (Chapter 4). In addition to these traditional instruments, there is a leading international initiative in this space called Biodiversity Understanding in Infrastructure and Landscape Development (BUILD), implemented by the Conservation Strategy Fund. BUILD aims to create lasting human capacity for energy and transport infrastructure analysis in partner countries that assesses the ecological and economic trade-offs involved in infrastructure investment decisions. It does this through a series of courses (e.g. in cost-benefit analysis, valuation methods, natural resource and environmental economics), regional forums, in-depth analyses of specific infrastructure projects, and information sharing. To date, BUILD has been applied in Peru, Bolivia and Brazil, as well as Africa and the Himalayan region (CSF, 2016).

6.6. Tourism

In 2015, Latin America received more than 96.6 million foreign visitors, the highest number reached over the last decade. The United States, and increasingly Latin American countries, are the main sources of visitors (UNWTO, 2016). Tourism offers significant economic opportunity; it represents the fourth largest export sector in Chile, with an estimated 4.5 million foreign visitors in 2015, and generates more than 8% of GDP in Mexico. It also presents an opportunity for increased biodiversity financing. However, tourism can also present a risk to biodiversity and ecosystems if it is not managed sustainably.

Nature-based tourism has particularly high potential in Latin America. While three-quarters of tourists in Chile visit a natural area, Brazil has yet to fully capitalise upon its natural wealth. A 2014 study estimated that the potential income in Brazil from tourism in protected areas could reach BRL 53 billion (USD 15 billion) over 10 years (Semeia, 2014). In 2013, just two national parks (Iguaçu and Tijuca) welcomed nearly 60% of visitors, and only 26 of Brazil's 68 national parks are open for tourism. Colombia is seeking to increase nature-based tourism by offering a 20-year income tax exemption for eco-tourism investments once they are certified by the Environment Ministry. The Colombian National Parks Authority is also implementing Community Ecotourism Programmes in some national protected areas that aim to improve the livelihoods of communities in the parks' zones of influence, while reducing pressures on natural resources by fostering sustainable economic activities (OECD/ECLAC, 2014).

Many governments have recognised the need to improve the sustainability of tourism to limit the negative environmental impacts of the sector's expansion, and to attract a growing number of environmentally-conscious, nature-seeking travellers. Chile and Mexico have developed national strategies for sustainable tourism and several countries are pursuing certification and labelling schemes. Colombia has a voluntary environmental certification system for tourism providers, while Mexico promotes eco-certification for tourism-related businesses in conjunction with the Rainforest Alliance and EarthCheck programmes (OECD, 2013). Chile has created New Sustainable Distinction Systems for Chilean Tourist Accommodation and Destinations based on global sustainable criteria suggested by the World Tourism Organisation (OECD and LEED, 2014).

6.7. Climate change

There are strong linkages between biodiversity and climate change adaptation and mitigation, creating opportunities to identify synergies in policy approaches that benefit both biodiversity and climate change goals. International climate change financing, market mechanisms and programmes may create opportunities to finance climate change adaptation or mitigation projects that also conserve or restore biodiversity. Climate change is expected to exacerbate water-related challenges, risks to biodiversity and the vulnerability of resource-based sectors in a number of regions in Latin America.

Forests in particular offer opportunities for synergistic climate-biodiversity benefits. Forests provide carbon sequestration services as well as species' habitats and other ecosystem services. Deforestation and forest degradation are also the second leading cause of climate change, responsible for about 15% of global GHG emissions. In some countries, such as Brazil, deforestation and forest degradation together are the main source of national GHG emissions (FCP, 2015). The UN Framework Convention on Climate Change (UNFCCC) process developed an initiative aimed at Reducing Emissions from Deforestation and Forest Degradation (REDD). In addition to stemming deforestation and forest degradation, the initiative also seeks to foster conservation, sustainable management of forests, and enhancement of forest carbon stock (REDD+). There are several international financing opportunities for countries able to demonstrate reduced GHG emissions through REDD+ activities (FCP, 2015). In 2014, Brazil was the first country to submit its forest reference emission level to receive payments under the REDD and REDD+ initiatives. It has received about half of the total approved international finance from REDD and REDD+ through its Amazon Fund (OECD, 2015b). Mexico has played a leading role in promoting the REDD+ initiative in international negotiations on climate change and Colombia has a Strategy for the Control of Deforestation and Forest Management which includes a REDD+ strategy. However, the benefits of REDD+ programmes on biodiversity may not be universal and could in some instances provide perverse incentives. For example, there may be poor overlap between biodiversity, carbon storage and the provision of other ecosystem services.

Several Latin American countries are working to leverage synergies between climate change mitigation, adaptation and biodiversity conservation. In its Intended Nationally Determined Contribution (INDC) under the UNFCCC, the government of Brazil set out goals to achieve zero illegal deforestation, compensate for GHG emissions from legal suppression of vegetation, and restore an additional 15 million ha of degraded pasturelands by 2030 (Government of Brazil, 2015). Chile has also committed to the sustainable development and recovery of 100 000 ha of mainly native forest land between 2020 and 2030 (Government of Chile, 2016). Mexico has developed a strategy for climate change adaptation in protected areas and its 2015 INDC embraces the concept of ecosystem-based climate change adaptation.[2] Colombia's national law and policy on climate change also recognises climate change as a driver of biodiversity loss and, in turn, the conservation and managing biodiversity as a strategy for climate change adaptation and the mitigation. In Peru, biodiversity-relevant sectors are at the core of the National Adaptation Plan and any of the actions taken to improve resilience are likely to also benefit biodiversity and ecosystems (Government of Peru, 2015).

Notes

[1] Silvopasture is an agroforestry practice that integrates livestock, forage production, and forestry on the same land-management unit.

[2] This includes reaching a rate of 0% deforestation by 2030, reforestation of watersheds and riparian zones, conserving and restoring ecosystems, strengthening the protection of priority species from the negative impacts of climate change, conservation and recovery of coastal and marine ecosystems, and integral management of water across agricultural, ecological, urban, industrial and domestic uses (Government of Mexico, 2015).

References

ABRASCO (2015), *Dossiê ABRASCO – Um Alerta Sobre os Impactos dos Agrotóxicos na Saúde*, Associação Brasileira de Saúde Coletiva, Escola Politécnica de Saúde Joaquim Venâncio, Rio de Janeiro/São Paulo, www.abrasco.org.br/dossieagrotoxicos/wp-content/uploads/2013/10/DossieAbrasco_2015_web.pdf.

Cevallos, D. (2006), *Latin America: Wave of Opposition Hits Hydroelectric Dams*, Inter Press Service News Agency, 6 May 2006, www.ipsnews.net/2006/05/latin-america-wave-of-opposition-hits-hydroelectric-dams/.

CIAP (2012), "Desaparición de las abejas y los residuos de plaguicidas en miel: Situación de la región de O'Higgins" [Disappearance of bees and pesticide residues in honey: Situation in the region of O'Higgins], *Agrocompetitivo*, Boletín N°1-2012, Centro de Investigaciones Aplicadas, Universidad Técnica Federico Santa María, Rancagua.

CSF (2016), *Biodiversity Understanding in Infrastructure and Landscape Development (BUILD)*, Conservation Strategy Fund, Washington D.C., http://conservation-strategy.org/en/node/1031#.U-EilPmSxHV.

Drutschinin, A. et al. (2015), Biodiversity and Development Co-operation, *OECD Development Co-operation Working Papers*, No. 21, OECD Publishing, Paris. http://dx.doi.org/10.1787/5js1sqkvts0v-en.

FAO (2015), *Global Forest Resources Assessment 2015*, Food and Agriculture Organisation, Rome, www.fao.org/forest-resources-assessment/en/.

FAO (2011), *Coastal Fisheries of Latin America and the Caribbean,* Food and Agriculture Organisation, Rome, www.fao.org/docrep/014/i1926e/i1926e.pdf.

FCP (2015), "What is REDD+?", Forest Carbon Partnership Facility, www.forestcarbonpartnership.org/what-redd.

Government of Brazil (2015), *Intended Nationally Determined Contribution Towards Achieving the Objective of the United Nations Framework Convention on Climate Change*, 28 September 2015, www4.unfccc.int/submissions/indc/Submission%20Pages/submissions.aspx.

Government of Chile (2016), *Intended Nationally Determined Contribution of Chile Towards the* Climate *Agreement of Paris 2015*, 2016-01-05, www4.unfccc.int/submissions/indc/Submission%20Pages/submissions.aspx.

Government of Colombia (2015), *Intended Nationally Determined Contribution*, 2015-09-07, www4.unfccc.int/submissions/indc/Submission%20Pages/submissions.aspx.

Government of Mexico (2015), *Intended Nationally Determined Contribution*, 2015-03-30, www4.unfccc.int/submissions/indc/Submission%20Pages/submissions.aspx.

Government of Peru (2015), *Intended Nationally Determined Contribution from the Republic of Peru*, 2015-09-28, www4.unfccc.int/submissions/indc/Submission%20Pages/submissions.aspx.

Huntley, B.J. and Redford, K.H. (2014), *Mainstreaming Biodiversity in Practice: a STAP Advisory Document*, Global Environment Facility, Washington, D.C., www.thegef.org/gef/sites/thegef.org/files/publication/Mainstreaming-Biodiversity-LowRes.pdf.

IDB (2015), *Biodiversity and Ecosystem Services Program: An Overview*, Inter-American Development Bank, Washington D.C., http://idbdocs.iadb.org/wsdocs/getdocument.aspx?docnum=38186826.

IEA (2016), *Oil Information 2016*, OECD Publishing, Paris, http://dx.doi.org/10.1787/oil-2016-en.

IEA (2015), *IEA Energy Atlas*, International Energy Agency, Paris, http://energyatlas.iea.org/.

Lizanco Rodriguez, A. et al. (2013), *The Great Debate: Mining in Latin America,* Colombia University Journal of International Affairs, 26 April 2013. http://jia.sipa.Colombia.edu/online-articles/great-debate-mining-in-latin-america/.

MMA (2015), Fifth National Report to the Convention on Biological Diversity, Ministry of the Environment, Brasília, www.cbd.int/doc/world/br/br-nr-05-en.pdf.

OECD (2015a), *Multi-dimensional Review of Peru: Volume 1.* Initial Assessment, OECD Development Pathways, OECD Publishing, Paris, http://dx.doi.org/10.1787/9789264243279-en.

OECD (2015b), *OECD Environmental Performance Reviews: Brazil 2015*, OECD Publishing, Paris, http://dx.doi.org/10.1787/9789264240094-en.

OECD (2013), *OECD Environmental Performance Reviews: Mexico 2013*, OECD Publishing, Paris, http://dx.doi.org/10.1787/9789264180109-en.

OECD DAC (2010), *Policy Statement on Integrating Biodiversity and Associated Ecosystem Services into Development Co-operation*, Development Assistance Committee, www.oecd.org/environment/environment-development/46024461.pdf.

OECD and LEED (2014), *Chile's Pathway to Green Growth: Measuring Progress at Local Level*, OECD Publishing, Paris, www.oecd.org/cfe/leed/Green_growth_Chile_Final2014.pdf.

OECD/ECLAC (2017), *OECD Environmental Performance Reviews: Peru 2017*, OECD Publishing, Paris, http://dx.doi.org/10.1787/9789264283138-en.

OECD/ECLAC (2016), *OECD Environmental Performance Reviews: Chile 2016*, OECD Publishing, Paris, http://dx.doi.org/10.1787/9789264252615-en.

OECD/ECLAC (2014), *OECD Environmental Performance Reviews: Colombia 2014*, OECD Publishing, Paris, http://dx.doi.org/10.1787/9789264208292-en.

Phillips, D. (2015), *Brazil's mining tragedy: was it a preventable disaster?* The Guardian, 25 November 2015, www.theguardian.com/sustainable-business/2015/nov/25/brazils-mining-tragedy-dam-preventable-disaster-samarco-vale-bhp-billiton.

Segal, S. (27 March 2014), "Chile among top twenty countries prone to environmental conflict", Santiago Times.

Semeia (2014), *Unidades de conservação no Brasil: a contribuição do uso público para desenvolvimento socioeconômico* (Protected areas in Brazil: the contribution of public use for socio-economic development), Instituto Semeia, São Paulo.

Tissot, R. (2012), *Latin America's Energy Future*, Discussion Paper No. IDB-DP-252, Inter-American Development Bank, Washington D.C., www.iadb.org/wmsfiles/products/publications/documents/37670565.pdf.

UN Water (2014), *Key Water Indicators Portal*, UN Water, Geneva, www.unwater.org/kwip.

UNWTO (2016), *UNWTO Tourism Highlights, 2016 Edition*, United Nations World Tourism Organization, Madrid, http://mkt.unwto.org/publication/unwto-tourism-highlights-2016-edition.

USGS (2016a) *2013 Minerals Yearbook – Brazil*, United States Geological Survey, March 2016, http://minerals.usgs.gov/minerals/pubs/country/2013/myb3-2013-br.pdf.

USGS (2016b), *2013 Minerals Yearbook – Mexico*, United States Geological Survey, May 2016, http://minerals.usgs.gov/minerals/pubs/country/2013/myb3-2013-mx.pdf.

USGS (2015a), *2013 Minerals Yearbook – Chile*, United States Geological Survey, March 2015, http://minerals.usgs.gov/minerals/pubs/country/2013/myb3-2013-ci.pdf.

USGS (2015b), *2013 Minerals Yearbook – Peru*, United States Geological Survey, December 2015, http://minerals.usgs.gov/minerals/pubs/country/2013/myb3-2013-pe.pdf.

Van Winkle, C. (2015), *Mainstreaming Biodiversity at the Sector Level: An Overview*, OECD Workshop on Biodiversity and Development, 18 February 2015, www.oecd.org/dac/environment-development/workshop-biodiversity-development-2015.htm.

World Bank (2013), *Future Looks Bright for Food Production in Latin America and Caribbean*, World Bank Group, Washington, D.C., www.worldbank.org/en/news/feature/2013/10/16/food-production-trade-latin-america-caribbean-future.

ORGANISATION FOR ECONOMIC CO-OPERATION AND DEVELOPMENT

The OECD is a unique forum where governments work together to address the economic, social and environmental challenges of globalisation. The OECD is also at the forefront of efforts to understand and to help governments respond to new developments and concerns, such as corporate governance, the information economy and the challenges of an ageing population. The Organisation provides a setting where governments can compare policy experiences, seek answers to common problems, identify good practice and work to co-ordinate domestic and international policies.

The OECD member countries are: Australia, Austria, Belgium, Canada, Chile, the Czech Republic, Denmark, Estonia, Finland, France, Germany, Greece, Hungary, Iceland, Ireland, Israel, Italy, Japan, Korea, Latvia, Lithuania, Luxembourg, Mexico, the Netherlands, New Zealand, Norway, Poland, Portugal, the Slovak Republic, Slovenia, Spain, Sweden, Switzerland, Turkey, the United Kingdom and the United States. The European Union takes part in the work of the OECD.

OECD Publishing disseminates widely the results of the Organisation's statistics gathering and research on economic, social and environmental issues, as well as the conventions, guidelines and standards agreed by its members.

www.ingramcontent.com/pod-product-compliance
Lightning Source LLC
Chambersburg PA
CBHW082354220526
45470CB00008B/2744